W9-ARU-347

# THE HARMONIOUS UNIVERSE

KEITH J. LAIDLER

# THE HARMONIOUS UNIVERSE

## THE BEAUTY AND UNITY OF SCIENTIFIC UNDERSTANDING

59 John Glenn Drive
Amherst, New York 14228-2197

Published 2004 by Prometheus Books



Inquiries should be addressed to
Prometheus Books
59 John Glenn Drive
Amherst, New York 14228-2197
VOICE: 716-691-0133, ext. 207
FAX: 716-564-2711
WWW.PROMETHEUSBOOKS.COM

08 07 06 05 04     5 4 3 2 1

Library of Congress Cataloging-in-Publication Data

Laidler, Keith James, 1916–2003
The harmonious universe : the beauty and unity of scientific understanding / Keith J. Laidler.
p. cm.
Includes bibliographical references and index.
ISBN 1-59102-187-1 (hardcover : alk. paper)
1. Science—Miscellanea. I. Title.

Q173.L25 2004
500—dc22

2004004009



Printed in the United States of America on acid-free paper

We have to admire in humility the beautiful harmony
of the structure of the world—as far as we can grasp it.

—Albert Einstein, 1945

# CONTENTS

# PREFACE

Today science is having a powerful influence on our material lives and is transforming our culture. For these reasons, it is useful for all of us to have a clear understanding of the methods of science and of how science influences the whole of society. That is not the same thing as saying that everyone should be given a formal scientific education; that would be impractical. After all, we can enjoy and appreciate music without being able to play or even read a note of it. So it is with science: today, with science and technology so dominating our lives, we should try to understand what science is about. In this book I try to help the general reader to do so.

The discussion herein is particularly directed to readers who have not studied much science. My intention is to illuminate a broad spectrum of topics that I think will help the general reader to gain some understanding of the methods of science and of the broad conclusions to which science leads. I have not hesitated to make some innocent simplifications, which I hope my fellow scientists will forgive; there is a wise maxim that says that a little inaccuracy can save a world of explanation. I have limited my discussions to those aspects of science that I think the nonscientist reader will find most interesting and helpful, and I have indicated some of the difficulties that scientists have had in reaching their conclusions. Some of the very latest developments are not dealt with, since they are usually too complex to be readily understood. Although the picture I have painted is a little impressionistic, it should give a reasonably correct idea of what we have learned from science.

I have tried to write as clearly as possible. It is a matter of some concern that certain books about science, although written very obscurely, have proved popular with the public. Perhaps some people think that an opaquely written book must be good one? Do members of the public really say to themselves, "I can't understand a word of this book; it must be a good one, and the author very clever"? I remember once, in my early days of teaching, being told by a student that the students in my class understood my lectures very well. Then she rather spoiled everything by adding, "But none of us can understand Professor X at all; but then, *he* is very brilliant." Professor X (not his real name, by the way) was not at all brilliant; he was rather confused about the basic concepts of science. Since I am convinced that there is no inverse correlation between obscurity and brilliance, I have striven in this book for correctness, cogency, and clarity.

I have not hesitated to include a few chemical equations and even a few mathematical ones—but all of them simple, with full explanations. I cannot believe that readers of this book are incapable of grasping the meaning of $E = mc^2$ or understanding an inverse-square law relationship. I have included diagrams wherever they seemed to be useful, as I have always liked them and have often found them to be a great help to a verbal explanation.

Keith J. Laidler
February 2003

# ACKNOWLEDGMENTS

I am greatly indebted to many people who have greatly helped me while I wrote this book. Dr. Alan Batten of the Herzberg Institute of Astrophysics, Dominion Astrophysical Observatory, gave me valuable advice on cosmological matters. Correspondence over the years with Dr. Walter Gratzer of the Randall Institute of King's College, London, has helped me with my understanding of molecular biology. Several people have read over all or part of the text at various stages and have made many constructive suggestions as to scientific content and style of presentation; I thank in particular Prof. John Holmes, Dr. John Morton, Dr. June Lindsey, and Dr. George Lindsey.

Since this book is primarily intended for readers who are not scientists, I needed help from people who would be willing to judge the book from that point of view and give me appropriate advice. I am particularly indebted to Dr. Christine Davenport, who has made a deep study of history and law but whose education included little science, for reading much of what I wrote in early drafts and advising me as to its suitability for the general reader. Many thanks are also due to my editor, Linda Regan, for her careful scrutiny of the book, and her always constructive criticisms.

My son Jim Laidler has prepared the line drawings in this book and has been of great help with finding suitable portraits. None of the portraits included is copyright-protected. For copies of the several portraits I am grateful to the Annenberg Rare Book and Manuscript Library, University of Pennsylvania, and in particular thank John Pollack for his personal help. For the interesting portrait of Ernest Rutherford in his early years I am especially indebted to Prof. John Campbell of the University of Christchurch, New Zealand, and to members of the Rutherford family.

#### A FEW
# POINTS ABOUT MATHEMATICS AND MEASUREMENT

In this book I have tried to keep the mathematics to a minimum and in particular have not used many mathematical equations. Some authors have eliminated equations completely from their books, and there is a popular mathematical theorem that sales of scientific books written for the public are halved for each equation included. I think that this is going a little too far. Some mathematical equations, like Einstein's $E = mc^2$ and Boltzmann's $S = k \log W$, can greatly help the reader, provided that they are properly explained. I have therefore not hesitated to use them as well as a few other equations.

The following are a few mathematical points of which some readers may need to be reminded, as well as a brief discussion of the metric system.

## SCIENTIFIC NOTATION

From time to time we have to use very large or very small numbers, and we can use what is called *scientific notation* to avoid writing out strings of zeros. Thus instead of writing "1,000,000," which is one million, we write "$10^6$," which is ten multiplied by itself six times, or one followed by six zeros. The superscript numbers are called *expo-*

*nents*. We can pronounce $10^6$ as "ten exponent 6," or "ten to the sixth power," or just "ten to the sixth." Similarly, a billion, or 1,000,000, 000, we write as $10^9$. If we want to write 602,200,000,000,000, 000,000,000 (which happens to be the number of molecules in 1 gram of hydrogen atoms), we write $6.022 \times 10^{23}$, which all will agree is far more convenient.

A similar notation applies to very small numbers. Suppose we are dealing with 1/1,000,000 (one millionth), which can also be written as 0.000001. To write it in scientific notation we note that one millionth is $1/10^6$, and write it as $10^{-6}$, which means that it is 1 divided by 10 six times. A simple way of deciding on the exponent in such a case is to see how far we have to move the decimal point to the right in order to get one; for 0.000001 we have to move it six places. The value of an important constant of nature called the Boltzmann constant is $1.381 \times 10^{-23}$, which means that it is 1.381 divided by 1 followed by 23 zeros.

An important point to note about numbers in scientific notation is that when the exponent changes by one, the number itself changes by a *factor* of ten. Thus $10^4$ is ten times as large as $10^3$, and $10^9$ (a billion) is a thousand times as big as $10^6$ (a million). The number $10^{-23}$ is a thousand times smaller than $10^{-20}$.

## LOGARITHMS

When we use *logarithms* we are doing the reverse of what we do in raising a number such as ten to a certain power. We have seen that for 1,000,000 we can write $10^6$. Conversely, we say that the logarithm of 1,000,000 is six. This kind of logarithm of a number is simply the power to which 10 is raised to get that number.

This particular logarithm is said to be to the "base ten," and it is called a *common logarithm*. The logarithm tables that were formerly used to make calculations but have now been largely superseded by electronic calculators and computers employed common logarithms. When using common logarithms it is useful, to avoid ambiguity, to use the notation $\log_{10}$; thus $\log_{10} 1{,}000{,}000 = 6$.

In scientific work we more often use what we call *natural logarithms*, written as $\log_e$ or more often as ln; for example, we write $\log_e 6$ or ln 6. These are to the base *e*, which is an odd but important little number that has the value 2.71828. . . . It is useful to note that natural logarithms are bigger than common logarithms by a factor of 2.303. Thus $\log_{10}$ 1,000,000 = 6, and ln 1,000,000 = 2.303 × 6 = 13.8.

In using logarithms it is important to remember that a change in a logarithm means a larger change in the quantity measured. The pH scale, commonly used by gardeners and others, is a common logarithmic scale, and it is an inverse measure of acidity. Thus a solution with a pH of 2 is ten times as acidic as one of pH 3 (ten times because it is a common logarithm, to the base ten). The Richter scale, named after American seismologist Charles Francis Richter (1900–85) is also a common logarithmic scale based on the amplitudes of vibrations that occur in an earthquake, as measured by a seismograph. In an earthquake that rates a 7 on the Richter scale, for example, the average amplitudes are measured to be ten times as large as for one that is a 6 on the Richter scale. The damage it causes will be much more than ten times as great.

## THE METRIC SYSTEM

Today the metric system is used universally by scientists, since many of the important relationships in science can be expressed and understood much more conveniently by its use. Officially, the metric system is today called the Système Internationale d'Unités (abbreviated SI). In this system, the base units of length, mass, volume, temperature, and time are, respectively, the meter, gram, liter, Celsius degree, and second. The metric system has become part of the language of science, and since it allows everyday calculations to be made much more easily, it has now been adopted for general use in most countries. Even in those few countries that have not adopted the metric system, primarily the United States and Burma, the metric system is necessarily used in the teaching of science.

In the present book the metric system is primarily used, usually with the Imperial (English) unit given in addition. The following tables of conversions may be useful to those not familiar with metric units.

### *Distance*

| | Inches | Feet | Yards | Miles |
|---|---|---|---|---|
| 1 meter (m) | 39.37 | 3.28 | 1.094 | |
| 1 centimeter (cm) | 0.3937 | 0.033 | | |
| 1 millimeter (mm) | 0.0394 | | | |
| 1 kilometer (km) | | | | 0.6214 |
| 1.609 km | | | | 1 |

### *Volume*

| | US pints | US quarts | US gallons |
|---|---|---|---|
| 1 liter (cubic decimeter, $dm^3$) | 2.11 | 1.057 | 0.264 |

1 Imperial gallon = 1.201 US gallons, with the same factor for the pint and the quart.

### *Speed*

| | Kilometer/hour | Feet/sec | Miles/hour |
|---|---|---|---|
| 1 meter/second | 3.6 | 3.28 | 2.237 |
| 1 kilometer/sec | 3,600 | 3,280 | 2,237 |
| 1 kilometer/hr | 1.000 | 0.911 | 0.6214 |

### *Mass*

| | Ounce | Pound | Long ton (2240 lb) | Short ton (2000 lb) |
|---|---|---|---|---|
| 1 gram | 0.0353 | 0.002205 | | |
| 1 kilogram | 35.27 | 2.205 | 0.00984 | 0.00110 |
| 1 tonne (metric ton = 1000 kg) | | 2,204.6 | 0.9842 | 1.102 |

### *Temperature*

The fundamental scale of temperature is the absolute or Kelvin scale, which is based on absolute zero. Some equivalents in the Celsius scale (very close to the centigrade scale) and the Fahrenheit scale are as follows.

| Kelvin (K) | Degrees Celsius (C) | Degrees Fahrenheit (F) |
|---|---|---|
| 0 K | –273.15 (by definition) | –459 |
| 3 K | –270.15 | –454 |
| 273.15 K | 0 | 32 |
| 293.15 K | 20 | 68 |
| 373.15 K | 100 | 212 |
| 5,000 K | 4,727 | 8,577 |
| 6,000 K (Sun's surface temperature) | 5,727 | 10,341 |
| 10,000 K | 9,727 | 17,540 |

### *Prefixes*

The following prefixes are commonly used with the various metric units:

| | | |
|---|---|---|
| tera (T) | = | 1 trillion = 1 million million ($10^{12}$) |
| giga (G) | = | 1 billion = 1000 million ($10^{9}$) |
| mega (M) | = | 1 million ($10^{6}$) |
| kilo (k) | = | 1 thousand ($10^{3}$) |
| centi (c) | = | 1 hundredth ($10^{-2}$) |
| milli (m) | = | 1 thousandth ($10^{-3}$) |
| micro (μ) | = | 1 millionth ($10^{-6}$) |
| nano (n) | = | 1 billionth = 1 thousand millionth ($10^{-9}$) |
| pico (p) | = | 1 trillionth = 1 million millionth ($10^{-12}$) |

# PRELUDE

Most of us would like to understand as much as we can about the universe around us. People who want to avoid going to much trouble are often willing to rely completely on the authority of others. Over the last few centuries, however, authority has been revealed to be a thoroughly unreliable route to accurate knowledge, much less to understanding. The thinkers of the distant past, even great ones like Aristotle, may have done their best with the information they had available, but they have almost always proved to be completely wrong. Great advances in science have now been made, and the only reliable way of understanding the universe is for us to look at the latest scientific evidence and draw our own conclusions. The experience of others is of course essential and helpful, but the final decision should be our own.

In Newton's time, about three centuries ago, the universe was thought to be a few thousand years old; we now know its age to be at least 12 billion years. Newton suspected the existence of atoms but knew little about them; we now know their sizes and many details of their structure. Life in all its different species was thought to have been created all at once, and few suspected that it has evolved, as we now know to be true.

Two centuries later, scientists knew much more. They knew that the universe was much older than previously thought and were beginning to learn something about the sizes and structures of atoms. It had become certain that life on Earth had evolved, but many details were obscure.

During the last century our knowledge of the universe grew tremendously. We now know that the universe is expanding and that it started with a dramatic cosmic event that we call the big bang. Today we can even make observations of the radiation left over from the big bang, just as we can look at fossils of early life forms. As for life, we now know many of the details of how it evolved, but much remains to be investigated. The main aim of this book is to present the most relevant modern scientific evidence that relates to the universe and to draw plausible conclusions from it.

In some respects, the universe is exceedingly complex. It is composed of about ninety different atoms, which in turn are made up of more fundamental particles. These atoms combine together into molecules, of which several million different ones have already been recorded, and there seems to be no limit to their number. Several different types of force hold molecules together, and the universe contains many different structures. Much of the universe is almost empty space, but a variety of structures exist within it—ordinary stars, neutron stars, red dwarfs, planets, and black holes, to name just a few.

Although the universe is very complicated, what we have learned about it does seem to fit into a relatively neat pattern, such that it possesses a certain harmony. Admittedly, in some respects the universe seems far from harmonious. One of the most striking things about it, in fact, is that it exhibits so much apparently discordant variety. We happen to live on a planet that suits our purposes reasonably well. The temperatures we experience allow us to survive for a time, and we have enough oxygen to breathe and enough light to see our way. But there are few, if indeed any, other parts of the universe where any kind of advanced life could exist. Most if not all of the rest of the universe is too hot, too cold, too bright, too dark, too dense, or too strongly invaded by lethal radiation, nor is it composed of the right chemical elements to support any but the most primitive forms of life.

In this sense, at least, the universe is far from uniform and not at all harmonious. But when we look at it more deeply, we find that the principles by which it has been formed are indeed uniform and harmonious. As we explore the laws of nature more carefully, we find that these principles are universal and absolute, applying everywhere in space and time. Powerful support for this is provided by astronomical observations. Light takes time to reach us, and anything we see at a very great distance is being seen as it was long ago; when we see the light from a star that is 10 billion light-years away, we are seeing the star as it was when the universe was young, but we have never yet seen any evidence that the laws of nature were at all different then. As far as we know, they are eternal.

The laws of nature are coherent in another sense: they do not depend in any way on the individual who studies them. This puts science in a class by itself; in all other intellectual studies, such as history or economics, the discipline is to some extent a function of the people who create it.

To make this book accessible to nonscientists I have necessarily given a simplified view of science. I have explained the relationships among the different branches of science, showing how together they lead to a unified conception of our universe and of living things within it. I explain some of the basic ideas and conclusions of chemistry, physics, astronomy, geology, and biology. In this way it is possible for the reader to see how evidence from all of these fields leads us to a picture of the formation and development of a harmonious universe and of life within it.

We begin with the microscopic universe—in other words, with the atoms, which are much too small to be seen with the naked eye. We will see how the evidence for the existence of atoms has revealed itself and will learn something about the fundamental particles of which they are composed. Then we will see very briefly how the ninety or so different kinds of atoms can combine together to form the several million different kinds of molecules that are already known to exist.

# 1

# THE MICROSCOPIC UNIVERSE

In June 1757, the great American scientist and statesman Benjamin Franklin (1706–90; see fig. 1) traveled on a ship from New York to England. His purpose was to petition King George II against repressive actions of the British government. While on the ship he had an interesting experience that had important scientific consequences. In his own words, written some years later,

> In 1757, being at sea in a fleet of 96 sail bound against Louisbourg [in Nova Scotia], I observed the wakes of two of the ships to be remarkably smooth, while all the others were ruffled by the wind, which blew freely. Being puzzled by the differing appearance, I at last pointed it out to our captain, and asked him the meaning of it? "The cooks," says he, "have, I suppose, been just emptying their greasy water through the scuppers, which has greased the sides of those ships a little"; and this answer he gave me with an air of some little contempt, as to a person ignorant of what every body else knew. In my own mind I at first slighted his solution, tho' I was not able to think of another. But recollecting what I had formerly read in PLINY, I resolved to make some experiment of the effect of oil on water, when I should have opportunity.

He seized this opportunity soon after his arrival in England, where he carried out a demonstration several times for the entertainment

**Figure 1.** Eminent American scientist and statesman Benjamin Franklin (1706–90) made significant contributions to the study of electricity and to many other aspects of science. The international respect in which he was held as a scientist led to the vital role he played in establishing the Constitution of the United States. (Edgar Fahs Smith Collection, University of Pennsylvania Library)

of some of his scientific friends. In 1774 he described one of these demonstrations as follows:

> At length being at CLAPHAM where there is, on the common, a large pond, which I observed to be one day very rough with the wind, I fetched out a cruet of oil, and dropt a little of it on the water. . . . I then went to the windward side, where [waves] began to form; and there the oil, though not more than a teaspoonful, produced an instant calm over a space several inches square, which spread amazingly, and extended itself gradually till it reached the lee side, making all that quarter of the pond, perhaps half an acre, as smooth as a looking glass.

The pond on Clapham Common, near London, where Franklin carried out this demonstration has been identified as Mount Pond, and it is still known by that name. Franklin went on to say that afterward, whenever he went into the country, he made a practice of taking with him "a little oil in the upper hollow joint of my bamboo cane, with which I might repeat the experiment as opportunity should offer; and I found it constantly to succeed."[1]

Walking sticks with removable heads were not uncommon at the time, being used, for example, by physicians for taking drugs to their patients. Franklin, who always liked a little joke, developed a trick of waving his stick like a magic wand over the water, thus releasing some oil, which stilled the waves as if by magic. One of the places where he demonstrated the effect of oil was Derwent Water in the Lake District. In 1773, a year before his paper was published, Franklin carried out a demonstration of wave-damping near Portsmouth, in the presence of some Fellows of the Royal Society.

It is useful to speculate on what Franklin did *not* proceed to do as a result of this experiment. As far as we know he never calculated the *thickness* of the layer of oil that he had produced, but we can make an estimate of it from his rough data. He said that a teaspoonful of oil produced a layer of half an acre, and if we perform the calculation the result is that the thickness is about 5 ten-billionths of a meter (roughly a billionth of an inch), which we write as $5 \times 10^{-10}$ m. Scientists today often use the angstrom unit for such small lengths; this unit has Å as its symbol and is equal to just 1 ten-billionth of a meter ($10^{-10}$ m). The thickness of Franklin's film was thus roughly 5 Å, which we know today to be typical of the size of a molecule; this is the name we give to the group of atoms combined together of which a chemical substance is composed. Many years later the distinguished scientist Lord Rayleigh (1842–1919) carried out more careful experiments on oil films and found about the same thickness.

Benjamin Franklin was one of the most remarkable men of all time. Born in 1706 in Boston, Massachusetts, he was the son of a chandler and soap boiler who had emigrated from England in 1683. He received hardly any education, and at the age of twelve he was indentured to a printer. From 1724 to 1726 he lived in England, where he became skilled in printing techniques, and on returning to Philadelphia he soon established his own publishing business, which was highly successful. In 1751 he was elected a member of the Pennsylvania House of Assembly and alderman of Philadelphia; from 1753 to 1774 he served as deputy postmaster general for the British colonies in North America. At the same time, over a period of years, he carried out a number of important scientific investigations.

Franklin was one of the few statesmen whose reputation was first gained through scientific achievements. His trip to England in 1757, in which he carried out his observations on oil films, was at the request of the Pennsylvania House of Assembly, and he made several subsequent diplomatic trips to Europe. With Thomas Jefferson and John Adams he was one of the three authors of the US Declaration of Independence, and in 1776 he was one of three commissioners sent to negotiate a treaty with France. In 1781 he played an active role in negotiating the final peace with Great Britain. After he had done so, one of his English friends, Christopher Baldwin, recalled to him his earlier experiments on oil films, and in a letter dated February 18, 1783, Baldwin wrote, "'Tis you who have . . . again poured the oil of Peace on the troubled Wave, and stiled [*sic*] the mighty Storm!"

As we have seen, in these oil experiments Franklin had done more than he realized; he had probably provided the first data from which the size of a molecule could be estimated. It may at first sight seem surprising that Franklin did not make the simple calculation of the thickness of the film, but it would probably never have occurred to anyone at that time. Atoms and molecules were still something of an abstraction, perhaps believed in but not considered to have much to do with one's experiments. When Franklin published his paper in 1774 he probably knew of an atomic theory that had been put forward in 1758 by the Croatian Jesuit priest Roger Joseph Boscovich (1711–87), who regarded atoms not as particles but as centers of force, of zero size. Thus, estimating the size of an atom or molecule might not then have seemed a reasonable thing to do.

There is yet another question that Franklin could have asked about his data but would never have thought of. But today we can ask, suppose, instead of just converting a small volume of oil into a thin layer, we imagine converting it into a filament, in which all the molecules are in a long string, all of them (except the ones at the end) touching their two immediate neighbors. In other words, we imagine converting the three-dimensional volume of oil in the teaspoon first into a two-dimensional layer, and then into a one-dimensional string. How long would the string be? Would it stretch across the Atlantic? Would it stretch to the Moon? People who do not know the answer and are invited to make a guess usually greatly underestimate the distance. The answer is that it would stretch to the Moon and back over ten thousand times—just from one teaspoonful of oil!

The surprising length that we have calculated comes from the smallness of the atoms. The cross-sectional area of the string of oil molecules is so tiny that the length has to be tremendously long. From the small size of the molecules, it follows that even a reasonably small amount of a substance must contain an enormous number of them. A liter (a little over 1 US quart) of water, for example, contains roughly $10^{25}$ molecules, each one of them consisting of two atoms of hydrogen and one of oxygen, so that we write it as $H_2O$. It is difficult for us to comprehend such a large number, or the tiny molecular size and mass that this number implies. One way to get some feeling for the number is to note that, as a matter of statistics, every time we inhale we take in over a million of the very same molecules once contained in the bodies of any famous person of old you want to mention—Aristotle or Julius Caesar. Every time we drink a glass of water, we can be confident that it includes many of the very same water molecules that were in the draft of hemlock with which Socrates ended his life.

Suppose that we connect from end to end the $10^{25}$ water molecules that are in a liter, so that we have a string of water molecules; it would stretch to the Moon and back over 10 million times.[2] A liter of water contains roughly a thousand times as many molecules as a teaspoonful of oil; thus its filament is about a thousand times as long. The length of the string produced from a liter of water is about 10 million million km ($10^{13}$ km), which is a little more than a light-year. (A light-year is the distance light travels in a year and is about $9.46 \times 10^{12}$ km.) The Moon is on the average "only" about 384,000 km (239,000 mi) from Earth, which is roughly 1 light-second. To and from the Moon is thus less than $10^6$ km, so a string $10^{13}$ km long can get there and back over $10^7$ (10 million) times.

No doubt some readers will think that I must have made a mistake in concluding that the molecules in a liter of water, if strung together, will reach to the Moon and back about 10 million times. It is appropriate that they do so; when dealing with science we must always be skeptical of anything we are told. I, too, did not believe the result of the calculation when I first made it. The following order-of-magnitude calculation may convince the skeptics that the conclusion is correct. The length of one unit in the chain of water molecules is substantially more than 1 angstrom, which is $10^{-10}$ meters. The length of a chain of $10^{25}$ water molecules is thus

$$10^{25} \text{ molecules} \times 10^{-10} \text{ m} = 10^{15} \text{ m} = 10^{12} \text{ km.}$$

The distance to the Moon, as stated above, is about 384,000 km (about 239,000 mi), so to the Moon and back is less than a million kilometers. The string $10^{12}$ km long will therefore get to the Moon and back more than $10^6$ (1 million) times. If we do the calculation more precisely we get about 10 million times.

We can get an idea of such an enormous number as $10^{25}$ in another way. Suppose we had that number of dollars and put it in a bank that gave no interest (lucky bank!). If we spent the money at the rate of a billion dollars a *second* (don't ask me how; I really have no idea, but it should be enjoyable!), how long would it be before we ran out of money? The answer is about a billion years. If the money were invested at only 1 percent per year, and we spent the interest at the rate of a billion dollars a second, the capital would accumulate, not diminish.

The following may also help us to gain some feeling for how small the atoms are. Suppose we look at a tiny ball bearing that is 1 millimeter (less than a twentieth of an inch) in diameter, and imagine magnifying it so much that each of the atoms becomes 1 millimeter in diameter. It is easy to calculate that the linear magnification required would be about 10 million and that the ball bearing itself would then have a diameter of 10 km (6.2 mi)! A glass of water with the same magnification would be about 100 km (62 mi) high, much higher than Mount Everest.

▲▲▲

When we look at a piece of metal it is natural at first to think of it as solid throughout. If we were to go on magnifying it indefinitely, would it continue to look the same? People thought for centuries that matter is completely solid, but that instinctive idea turned out to be wrong. A few of the ancient philosophers put forward the theory that matter was made up of atoms, notably the Latin poet Titus Lucretius Carus (c. 99–55 BCE). Isaac Newton also assumed the existence of atoms, partly to explain how light could pass through certain materials such as glass; how could it do so unless there were spaces between the constituents of matter?

Until the latter part of the nineteenth century, most scientific investigators thought of atoms as if they were little hard balls. It was believed that they carried hooks of some kind, because atoms had to hang on to other atoms in some way in order to form molecules. Thus a water molecule, $H_2O$, was thought of as an oxygen atom having two hooks, with each one of them attached to a hook on a hydrogen atom, which carries only one hook. Scientists did not actually call them hooks; they called them valences (US) or valencies (UK), but the idea was the same. Oxygen was said to have a valence of 2, hydrogen a valence of 1. A pair of hooks holding two atoms together was called a valence bond, or just a bond.

During the nineteenth century, much work was done on the study of electricity, which led to deeper understanding of the nature of atoms. The great scientist Michael Faraday (1791–1865), whom we shall meet often in this book, made many advances in the study of electricity, and one of his investigations related to *electrolysis*, in which an electric current passing through a solution decomposes a chemical compound into its component parts. When electricity passes through water (preferably with something in it just to make it conduct electricity better), the water is split into hydrogen and oxygen. These gases appear at the two poles, or electrodes, that are put into the liquid in order for the current to pass. These experiments of Faraday soon led to the suggestion that there must be some electric particle that passes along a wire when a current passes through it. This particle must be able to add on to atoms, and under some circumstances atoms must be able to give up the particle to the electrodes. The details of some of this are a little complicated, and we need not go into them. The main point is that this work produced evidence of the existence of a particle of electricity that passes along wires and somehow interacts with molecules. In 1874 Irish physicist George Johnstone Stoney (1826–1911) discussed the problem in a way that explained many things. He suggested that there was a negative particle of electricity, to which he gave the name *electron*, a name that has stuck. He even estimated the value of the charge on the electron but had no way of estimating its mass.

Further insight into the nature of electrons was obtained from experiments of a completely different kind. They involved discharge tubes, with which we are all familiar in the form of fluorescent lights, like the neon signs that are often used for advertising. A discharge tube is essentially a tube from which as much air as possible has been pumped out and through which electricity is passed. The bright discharge obtained is partly caused by a flow of the negatively charged electrons In the last decade of the nineteenth century, British physicist Joseph John Thomson (1856–1940; see fig. 2) began to study them in detail. Through carefully designed experiments he was able to measure both the mass and the charge of an electron, and in 1897 he announced that an electron was about two thousand times lighter than a hydrogen atom. He went on to investigate positively charged particles, which he found to be much heavier, their masses being more or less the same as the masses of atoms.

It soon became clear that an atom is nothing like a tiny hard ball. Ernest Rutherford (1871–1937; see fig. 3) deduced that an atom has a nucleus, which is very much smaller than the atom itself; a nucleus occupies only about 1 trillionth ($10^{-12}$) of the volume of the atom. Rutherford, who later became Baron Rutherford of Nelson, is recognized as the chief founder of the science of nuclear physics. He was

**Figure 2.** Joseph John Thomson (1856–1940), Cavendish Professor at Cambridge for many years, is famous for first measuring some properties of the electron.

born near Nelson, New Zealand, and was a student at Canterbury College, Christchurch, New Zealand. In 1895 he went to Cambridge to work under Thomson on radioactive processes, and in 1898 he became professor of physics at McGill University in Montreal, returning to Britain in 1907 to become professor at Manchester University.

In 1896 French physicist Antoine Henri Becquerel (1852–1908) discovered radioactivity, and two years later Rutherford observed that radioactive substances emit certain kinds of radiation. The less penetrating rays he called alpha ($\alpha$-) rays, which were later identified as the nuclei of helium atoms. The more penetrating ones he called beta ($\beta$-) rays, which were found to be electrons. In 1900 he observed the emission of even more penetrating rays, which he called gamma

**Figure 3.** Ernest Rutherford (1871–1937), First Baron Rutherford of Nelson, is recognized as the founder of the science of nuclear physics. He made many pioneering investigations on radioactivity and on atomic nuclei. (A photograph taken at McGill University, where Rutherford did most of the work that led to his Nobel Prize)

(γ-) rays. These turned out not to be beams of particles, but rather similar to light but much more energetic. In 1903 Rutherford recognized that radioactive decay often occurs by successive transformations and that elements sometimes changed their chemical identity. This idea that one chemical element could be converted into another had been suggested by early alchemists, who called it *transmutation*, but they had been discredited. Rutherford's idea that it could really occur was therefore at the time a revolutionary one, and he once shouted to a colleague, "For Mike's sake, Soddy, don't call it transmutation. They'll have our heads off as alchemists."

The evidence for the fact that nuclei are tiny compared with the atoms themselves came from experiments carried out in 1907, in Rutherford's laboratory at Manchester University. His research students Hans Geiger (1882–1945) and Ernest Marsden (1889–1970) fired alpha particles (α-particles), the positively charged nuclei of helium atoms, at a thin sheet of metal foil and determined the direction in which they were scattered. Most of the particles were scattered by only a few degrees, but about 1 in 8,000 of them bounced right back from the foil. In Rutherford's words, "It was as though you had fired a fifteen-inch shell at a piece of tissue paper and it came back and hit you." From this result, Rutherford deduced that there was powerful electric repulsion between the positively charged α-particle and a positively charged nucleus within the metal, and he was thus able to estimate the size of the nucleus. To gain some idea of the small size of the nucleus of an atom compared with the size of the atom itself, imagine an atom magnified to the size of a large cathedral. The radius of the nucleus would be about the size of a golf ball. If a nucleus were magnified so that its radius is about the width of this page, the electrons would be more than several city blocks away. The size of the atoms results from the motion of the electrons associated with the nuclei.

After World War I, Rutherford observed that some nuclei could be made to disintegrate by bombarding them with α-particles. In 1908 he was awarded the Nobel Prize for chemistry. This surprised him, as he thought he was a physicist, and he commented that no transformation he knew of was faster than his change from a physicist into a chemist. It is curious that Rutherford never received a second Nobel Prize for his work on the nuclear atom, which was of great importance. Rutherford's genius was aptly summed up by the discoverer of the neutron, Sir James Chadwick, who said that Rutherford's mind was like the bow of a battleship—it had no need to be as sharp as a razor since it had so much weight. Rutherford was an extremely good-natured and boisterous extrovert, never noted for his modesty. In 1919 he succeeded Thomson as Cavendish Professor at Cambridge.

It took some years to sort out the question of just how the nuclei of the individual chemical elements are constructed. An important clue had been provided as early as the 1860s by Russian chemist Dmitry Ivanovich Mendeleyev (1834–1907). In the course of writing a book on chemistry he prepared a set of cards, one for each of the known chemical elements, of which about sixty had been identified at the time. On each of the cards he listed the chemical properties of the element. As a form of relaxation he enjoyed playing the card game called patience or solitaire, and on laying down his cards for the chemical elements he was struck by the fact that if the cards were dealt in rows of certain lengths, in order of increasing relative atomic mass, elements with similar chemical properties often lay in the same column. There were a few exceptions, but Mendeleyev boldly assumed that their atomic masses were in error (a conclusion that in fact turned out to be wrong; there was a more subtle reason for the anomalies). The table he formulated from these cards became known as the *periodic table*, and the number of an element when all the elements are listed in order of increasing atomic mass became known as its *atomic number*. There were some gaps in the table, but Mendeleyev correctly predicted that they corresponded to elements that had not yet been discovered. His suggestions were at first ridiculed, but as time went on the missing elements were discovered and found to have the properties he had predicted. Mendeleyev might have been awarded a Nobel Prize for what is now regarded as a great contribution to science. However, the work had been done considerably before 1901, and Alfred Nobel had specified in his will that the prizes were to be awarded for work done in the previous year.[3] That is probably why Mendeleyev never received a Nobel Prize, but in 1955 the newly discovered element with an atomic number of 101 was named mendelevium (Md) in his honor.

In 1913 a contribution of fundamental importance was made by Henry Gwyn Jeffreys Moseley (1887–1915), who was working in Rutherford's laboratories at the University of Manchester. Moseley measured the characteristic frequencies the x-ray spectral lines of numerous elements, discovering a relationship from which he was able to deduce the charge of the atomic nucleus; this is the number of protons in the nucleus, equal to the number of orbiting electrons in the neutral atom. This number turned out to be the same as the atomic number, which as we have seen is more or less the ordinal number of an element when the elements are all listed in order of increasing atomic weight. Moseley's brilliant career was cut short at the age of twenty-seven, when he was killed in action at Gallipoli in World War I.

At the time of Moseley's work, scientists thought that atomic nuclei were composed of protons and a smaller number of electrons that have an equal and opposite charge and would neutralize some of them. However, British physicist James Chadwick (1891– 1974) thought it unlikely that a proton and an electron could remain close together and concluded that they would instead at once form a new particle, which in anticipation of its discovery he called a *neutron*. For some twelve years Chadwick looked for experimental evidence for the existence of the neutron and in 1932 carried out experiments that proved their existence beyond doubt. Although neutrons can now be detected reasonably easily, they had escaped detection for some years because the earlier techniques were inadequate. Since they bear no electric charge, neutrons can pass right through matter without doing much to it and therefore without creating much of a disturbance—just as an invisible man whose mass was concentrated in a ball 1 million millionth of his normal size might escape notice in a crowd. Neutrons can make themselves felt, however, if they are caused to move at high speed by the use of a powerful accelerator such as a cyclotron. By the use of modern techniques physicists can generate and experiment with neutrons, so that we now know a good deal about them. When combined with protons in an atomic nucleus, neutrons can be quite stable, but by themselves they are very unstable, living for only about fifteen minutes.

▲▲▲

The simplest atom is the hydrogen atom (see fig. 4, *a*), in which there is one electron bearing a charge that exactly balances that of the nucleus, such that the atom as a whole has no charge. This nucleus that has a single positive charge is called a *proton*, and it is one of the fundamental particles of nature. We can call this particular form hydrogen-1, or *protium*, and give it the symbol H. It is also helpful to write it as $^{1}_{1}H$, the superscript 1 meaning that there is one particle in the nucleus, a single proton, and the subscript 1 meaning that the charge is one. About 90 percent of all the atoms in the universe are of this type, and although they are the lightest of all atoms,

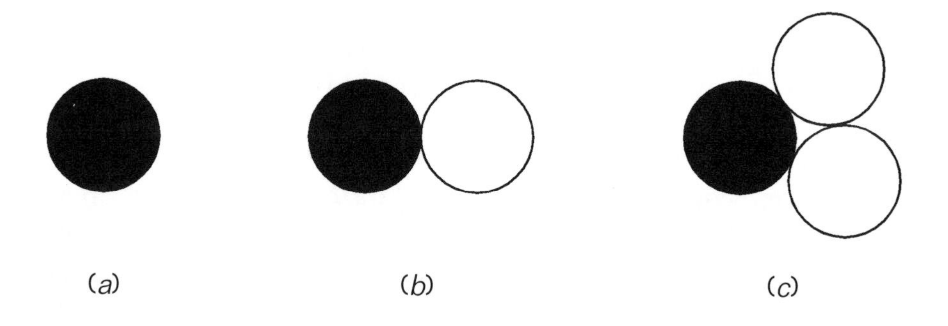


**Figure 4.** The nuclei of the three isotopes of hydrogen, showing the numbers of protons (filled circles) and neutrons (open circles): *a*, ordinary hydrogen, or protium, whose nucleus is a proton that has an orbital electron; *b*, deuterium, in which the nucleus consists of a proton and a neutron; *c*, tritium, in which the nucleus consists of a proton and two neutrons. In the neutral atoms, each of these nuclei has a single electron associated with it. According to the old theory, this electron is regarded as being in an orbit, but according to quantum mechanics it exists as a kind of cloud, referred to as an *orbital* (see chap. 4).

they account for about three-quarters of the mass of the entire universe.

There are two other forms of hydrogen. One of them differs from the ordinary hydrogen atom H in that the nucleus is about twice as heavy. The reason for this is that the nucleus consists of a proton and a neutron, which has almost exactly the same mass as the proton but no electric charge. This particular form of hydrogen we call hydrogen-2, or *deuterium*, and give it the symbol D (see fig. 4, *b*). We can also write it as $^{2}_{1}H$, the superscript 2 meaning that there are two particles in the nucleus and the subscript 1 meaning that the charge is one, the charge of the single proton. The well-known substance *heavy water*, used in some nuclear reactors, is made up of an oxygen atom combined with two deuterium atoms, written as $D_2O$. The D isotope is much less abundant than H: the relative abundance of deuterium ($^{2}_{1}H$) is only 0.015 percent that of ordinary hydrogen ($^{1}_{1}H$), being 99.985 percent.

There is yet another possibility for the hydrogen atom. A proton can also become attached to two neutrons, such that its mass is three times as great as that of $^{1}_{1}H$. We call this form hydrogen-3, or *tritium*, and give it the symbol T or $^{3}_{1}H$ (see fig. 4, *c*). This form of hydrogen is very unstable and does not survive very long, so its abundance in the universe is essentially zero. The word *isotope* is used to refer to atoms whose nuclei have the same number of protons but differ in the number of neutrons. Different isotopes of an element are always very similar in chemical and physical properties, which depend much more on the charge on the nucleus than on the mass.

The nuclei of all atoms are made up of protons and neutrons, but only certain combinations give rise to reasonably stable nuclei; others cause the nucleus to be unstable and to break up. For example, there is no nucleus composed of just two protons. There is one, however, that contains two protons and one neutron (see fig. 5, *a*), and another that contains two protons and two neutrons (see fig. 5, *b*). To make a neutral atom out of each of these, we must add two electrons. We now no longer have hydrogen but another chemical element, since the chemical behavior of an atom is determined by the number of protons associated with the nucleus. This atom that has two protons in its nucleus is called *helium*. The two forms of helium, which we write as $^{3}_{2}He$ and $^{4}_{2}He$, are isotopes of helium (remember that the subscript tells us the number of protons, the superscript the total number of particles in the nucleus, called *nucleons*). Since the proton and the neutron weigh much the same, $^{4}_{2}He$ is about four times as heavy as an ordinary hydrogen atom, whereas $^{3}_{2}He$ is three times as heavy. Helium found in nature is almost all $^{4}_{2}He$, the relative abundance of this form being 99.99986 percent, so the amount of $^{3}_{2}He$ is exceedingly small. Of all the atoms in the universe, about 10 percent are $^{4}_{2}He$, and because these atoms are four times heavier than the more abundant hydrogen atoms, they account for about a quarter of the total mass contributed by the ordinary matter in the universe.

The element carbon, which plays so important a part in living systems, exists in three known isotopic forms. The most common, with a relative abundance of 98.9 percent, is $^{12}_{6}C$, or carbon-12, its nucleus consisting of six protons and six neutrons, so that the mass number is 12. There is also $^{13}_{6}C$, carbon-13, in which there is an extra

**Figure 5.** The nuclei of the two known isotopes of helium. The one with two neutrons, $^{4}_{2}He$ (*b*), is by far the more common. An interesting fact, significant in relation to the origin of the chemical elements (see chap. 8), is that isotopes $^{2}_{2}He$ and $^{5}_{2}He$ do not exist. The former would have two protons and no neutrons in its nucleus, which would be highly unstable. The latter would have two protons and three neutrons, which again is too unstable an arrangement.

neutron; the relative abundance of it is 1.1 percent. A third form, $^{14}_{6}C$, has also been detected in tiny amounts.

For the lighter elements, it is often the case that the isotopes in which the numbers of protons and neutrons are the same is the most common. This is true for oxygen, where the most common isotope is $^{16}_{8}O$, with 8 protons and 8 neutrons. It is also true for nitrogen, where the common form is $^{14}_{7}N$, with 7 protons and 7 neutrons. With the heavier elements, however, this is no longer the case. Instead, the only isotopes that are at all stable tend to have many more neutrons than protons. An example is

Gold $^{197}_{79}Au$ 79 protons and 118 neutrons—197 nucleons.

Gold has no isotopes, which is unusual for nuclei containing so many particles. Platinum has several isotopes, of which the most abundant, with a relative abundance of 33.8 percent, is

Platinum $^{195}_{78}Pt$ 78 protons and 117 neutrons—195 nucleons.

The elements just mentioned, gold and platinum, are quite stable, but the nuclei of many of the heavier elements tend to break up spontaneously and are therefore said to be *radioactive*. Here are some examples of radioactive nuclei:

| | | |
|---|---|---|
| Polonium | $^{209}_{84}Po$ | 84 protons and 125 neutrons—209 nucleons; |
| Radium | $^{226}_{88}Ra$ | 88 protons and 138 neutrons—226 nucleons; |
| Thorium | $^{232}_{90}Th$ | 90 protons and 142 neutrons—232 nucleons; |
| Uranium | $^{238}_{92}U$ | 92 protons and 146 neutrons—238 nucleons. |

In the case of polonium, radium, and thorium, only one isotope has been detected. With uranium, on the other hand, three isotopes have been detected, their relative abundances being

| | | |
|---|---|---|
| $^{234}_{92}U$ | uranium-234 | 0.006 percent; |
| $^{235}_{92}U$ | uranium-235 | 0.720 percent; |
| $^{238}_{92}U$ | uranium-238 | 99.274 percent. |

Becquerel, discovering radioactivity in 1896, made his great contribution by mistake and never properly understood what radioactivity really was. He was a member of a remarkable scientific dynasty in that he, his grandfather, his father, and his son all became directors of the Musée d'histoire naturelle in Paris; when his son Jean Becquerel retired from that position in 1948, the museum did not have a Becquerel as director for the first time in 110 years. And when Jean died in 1953, the august and exclusive Académie des sciences ceased to have a Becquerel as a member for the first time in well over a century.

Henri Becquerel was looking for something else when he accidentally discovered what later was called radioactivity. Certain minerals and other solids emit radiation when light is shone on them and are said to *fluoresce*; often they cease to emit as soon as the light is turned off, but if the radiation persists for as long as a few seconds or minutes after the light is turned off, we say that they *phosphoresce*. In the course of an investigation of fluorescence, Becquerel happened to put aside, in a drawer, a package containing a uranium salt and a photographic plate, wrapped in black paper to prevent exposure of the plate.

After they had been there a few days, he for some reason developed the plate; why he did this is something of a mystery, since he firmly believed that there would be no blackening. It is fortunate that he did so, for he discovered that there was blackening. He continued investigations along these lines, but failed to gain any real understanding of it, thinking that the effect was some unusual kind of fluorescence, never realizing that the effect was specifically due to the uranium in the ore. It was left for Pierre and Marie Curie to discover the true origin of the emissions and for Rutherford to identify their nature.

Manya Sklodowska, who is now remembered as Marie Curie (1887–1934; see fig. 6), was born in Warsaw, Poland, a country then dominated by Russia. In 1891 she went to Paris to study and carry out research in physics. She married Pierre Curie (1859–1906) in 1895 and carried out research with him in his laboratories at the Municipal School of Industrial Physics and Chemistry in Paris. His main interests were then electricity and magnetism, and one of his achievements was to design a highly sensitive electrometer to measure electricity. After Becquerel's discovery in 1896, they decided to concentrate on similar investigations to his. Marie soon realized that the phenomenon was not one of fluorescence but was specifically due to the uranium in the ore. She then found that substances containing the element thorium emitted the same kind of radiation as did uranium. She examined a number of ores, one of which was pitchblende, which she found to emit much more powerful radiation than could be accounted for by the amounts of uranium and thorium it contained. She concluded that it must contain some other elements that emitted radiation. Marie and Pierre Curie then embarked on a lengthy and laborious study in which they extracted, from a large amount of pitchblende, tiny quantities of other materials that were extremely powerful emitters of radiation—in other words, were highly *radioactive*, a word that they introduced in the course of their work.

In 1898 Marie and Pierre Curie announced the discovery, as a minor ingredient of pitchblende, of an element they named polonium, in honor of Marie's native country. They found that polonium was four hundred times more radioactive than uranium. Later in the same year they discovered the element radium and showed it to be

**Figure 6.** Marie Curie (1867–1934), famous for her discovery of the elements polonium and radium and for her investigations of radioactivity. (Edgar Fahs Smith Collection, University of Pennsylvania Library)

nine hundred times more active than uranium. They then undertook the highly laborious task of obtaining a larger sample of a radium compound. By 1902 they had obtained, from many tons of a rich ore, a tenth of a gram of radium chloride. It was intensely radioactive, giving out heat and glowing in the dark; the radiation from it decomposed water into hydrogen and oxygen. Since the harmful effects of radiation were not then known, they took no precautions, and afterward their health was seriously affected.

In 1903 the Curies shared with Henri Becquerel the Nobel Prize for physics. Pierre was killed in a street accident in 1906; he had been made professor of physics at the Sorbonne, and Marie succeeded to the position. She continued her work on radioactivity and in 1911 was awarded a second Nobel Prize, in chemistry, specifically for her discovery of polonium and radium.

It was Ernest Rutherford who, with some of his colleagues such as Frederick Soddy (1877–1956), first discovered just what was happening in a radioactive *disintegration,* as a radioactive process came to be called. Uranium is the most abundant radioactive element in nature, and Rutherford found that the radiation it emits takes the form of the nuclei of $^{4}_{2}He$ atoms, introduced earlier as α-particles. To see what happens we can do some arithmetic as follows:

$^{238}_{92}U$ (92 protons + 146 neutrons) → $^{4}_{2}He$ (2 protons + 2 neutrons) + ?

The product must contain 92 – 2 = 90 protons, which means that it is the element thorium (remember that the name of the element is determined by the number of protons in the nucleus). It has 144 neutrons, the total number of particles being 234:

$^{238}_{92}U$ (92 protons + 146 neutrons) → $^{4}_{2}He$ (2 protons + 2 neutrons) + $^{234}_{90}Th$ (90 protons + 144 neutrons).

The only isotope of thorium that has any stability is $^{232}_{90}Th$, but the form produced in this process has an additional two neutrons. It also undergoes radioactive disintegration but not in the same way as uranium $^{238}_{92}U$. Instead of giving off an α-particle, which is a helium nucleus, Rutherford found that it emits a β-particle, which is an electron. Emission of an electron is equivalent to converting a neutron into a proton:

neutron → proton + electron.

When a $^{234}_{90}Th$ (90 protons + 144 neutrons) nucleus emits a β-particle, the nucleus of the product therefore has 91 protons and 143 neutrons, the total number remaining at 234. We can thus write the process in terms of the balanced equation

$^{234}_{90}Th$ (90 protons + 144 neutrons) → electron + $^{234}_{91}Pa$ (91 protons + 143 neutrons).

The element that has a nuclear charge of 91 is called protoactinium and given the symbol Pa. We can write the electron, or β-particle, as ${}_{-1}^{0}\beta$, the subscript −1 meaning that its charge is −1; the superscript 0 means that the tiny mass is effectively zero. Thus we write the disintegration as

$$ {}_{90}^{234}\mathrm{Th} \rightarrow {}_{-1}^{0}\beta + {}_{91}^{234}\mathrm{Pa}. $$

These radioactive processes are typical of many other spontaneous nuclear processes that occur. They differ in that in one of them a helium nucleus (α-particle) was emitted, in the other an electron. They also differ considerably in another way, namely the rates at which the disintegrations occur. Rutherford found that when a nucleus undergoes radioactive disintegration, it does so in such a way that however much of it we take at a given time, half of that amount will have disappeared after a certain time (see fig. 7). This time is known as the *half-life,* and it has a characteristic value for each nucleus. An interesting feature of the half-life for a radioactive process is that it is completely independent of outside factors, such as temperature or pressure.

The half-life of ordinary uranium, ${}_{92}^{238}\mathrm{U}$, is about 4.5 billion years. In other words, if we have ten grams of it now, we will have five grams of it in about 4.5 billion years; the rest will have decomposed into thorium and helium by the process written above. The isotope of thorium produced in the disintegration of uranium, ${}_{90}^{234}\mathrm{Th}$, has a much shorter half-life, only 24.1 days. There is a vast range of half-lives for radioactive substances. The common form of radium, ${}_{88}^{226}\mathrm{Ra}$, has a half-life of 1,620 years. Some isotopes have half-lives of only a small fraction of a second and are therefore very difficult to study. An isolated neutron disintegrates into a proton and an electron (a β-particle) with a half-life of 15.3 minutes.

▲▲▲

So far we have said hardly anything about how atoms join together to form molecules, which is something with which the science of chemistry is particularly concerned. The first serious efforts to under-

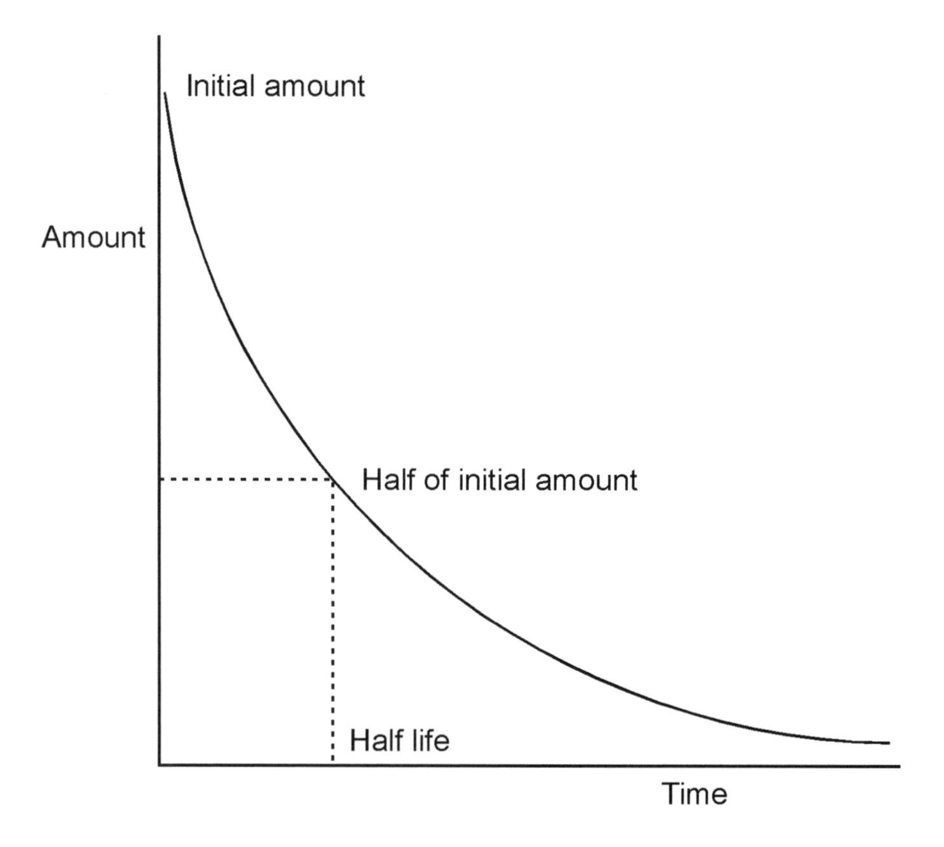


**Figure 7.** The amount of a radioactive substance remaining, plotted against time. The important feature of the relationship is that however much we start with, we will have half of that amount in a fixed period of time, known as the *half-life* for the particular nucleus.

stand the structures of molecules were made in the middle of the nineteenth century. Previously, chemistry had been largely an empirical science, without too much theory; chemists prepared new chemical compounds, determined how many of the various kinds of atoms each molecule of them contained, and discovered how the substances reacted with one another. From these studies, inferences could be drawn about the way in which the atoms were connected together in the molecules. Many of the inferences proved to be correct, but uncertainties remained until x-ray methods and other physical methods could be used.

An important advance was made in 1852 by English chemist Edward Frankland (1825–99), who had received some of his training in chemistry in the laboratories of Robert Bunsen (1811–99) at the University of Marburg, where he obtained his doctorate. In 1852 Frankland drew attention to the fact that particular chemical elements tended to combine with a fixed number of other atoms. Nitrogen, phosphorus, and arsenic, he pointed out, tend to join on to three atoms, as in the compounds $NH_3$, $NI_3$, $PH_3$, $PCl_3$, $AsH_3$, and $AsCl_3$. He commented, "No matter what the character of the uniting atoms may be, the combining power of the attracting element, if I may be allowed the term, is always satisfied by the same number of atoms." In about 1865, the word *valence* or *valency*, from the Latin *valencia* 'power'—compare the word *equivalence*—began to be used for what Frankland had called the "combining power." He realized that an atom could have two or more different valences; phosphorus, for example, can form both $PCl_3$ and $PCl_5$.

The vast majority of chemical compounds are what we call organic compounds, since they are of the type found in living organisms. The official definition today of an organic compound is simply that its molecule contains a carbon atom. Sometimes this produces apparent anomalies: calcium carbide, for example, is an organic compound according to this definition, although it has no connection with living systems. During the nineteenth century, chemists devoted much effort to establishing the structures of some of the vast number of chemical compounds that exist.

German chemist Friedrich August Kekulé (1829–96) played an important role in advancing our ideas about the structures of organic molecules. Much credit should also go to Archibald Scott Couper (1831–92), a Scotsman who went to work in Paris in 1856. In 1858 he published a paper that gave many more correct structures than were given by Kekulé at the same time. Unfortunately, he began to develop serious psychological problems and after his return to Scotland spent periods of time in a mental institution; his work is almost forgotten.

Kekulé was born in Darmstadt and attended the University of Giessen, where he studied under the great chemist Justus von Liebig (1803–73). He later went to Paris and undertook further studies with distinguished chemist Jean-Baptiste Dumas (1800–84). As a result of the work of Couper, Kekulé, and others, the structures of the compounds of carbon began to be understood. Chemists found that a carbon atom tends to combine with four atoms of certain elements such as hydrogen and chlorine, which have a valence of one. It will also combine with two atoms of oxygen to form carbon dioxide ($CO_2$); the oxygen atom has a valence of two. In modern notation some simple compounds can be represented in two dimensions, as

$$\begin{array}{ccccc} & & H & & \\ & & | & & \\ H & - & C & - & H \\ & & | & & \\ & & H & & \end{array} \qquad \begin{array}{ccccc} & & Cl & & \\ & & | & & \\ Cl & - & C & - & Cl \\ & & | & & \\ & & Cl & & \end{array} \qquad O{=}C{=}O.$$

Four valence bonds emanate from each carbon atom, which has a valence of four, and one from each hydrogen or chlorine atom. which have valences of one. The oxygen atom, on the other hand, has a valence of two; two bonds (called a double bond) therefore connect it to a carbon atom.

There are many organic molecules in which carbon atoms are joined together, sometimes in long strings. A simple example is ethane, in which there are two carbon atoms and six hydrogen atoms. One valence bond of each carbon atom is used to bind two carbon atoms together, whereas the other bonds are used to attach six hydrogen atoms as follows:

$$\begin{array}{ccccccc} & & H & & H & & \\ & & | & & | & & \\ H & - & C & - & C & - & H \\ & & | & & | & & \\ & & H & & H. & & \end{array}$$

Kekulé became professor of chemistry at the University of Ghent in 1858, and in 1867 he moved to Bonn. By introducing the idea that a chain of carbon atoms can be linked together, Kekulé was led to suggest a structure for the benzene molecule. This substance had been discovered in 1825 by Michael Faraday, who rather surprisingly had found it in oils rather than in the many aromatic spices and resins in which it is commonly present. Benzene is of particular importance in forming the basis for a vast number of other compounds of carbon. Compounds that contain structures like benzene are called *aromatic compounds* because they often have a characteristic spicy and pleasantly pungent smell.

The structure of benzene presented chemists with quite a difficult problem. It is easy to suggest a large number of structures that satisfy the valence rules (hydrogen being univalent and carbon tetravalent), but some of them seemed inconsistent with the properties of the substance, and how is one to decide between the different possibilities? One structure suggested was

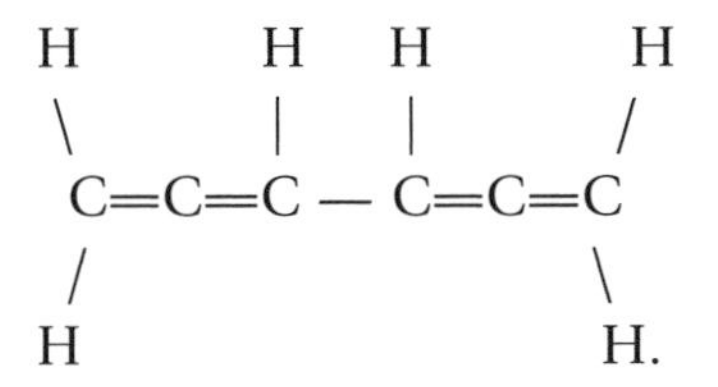


It turned out, however, to be impossible to explain the properties of benzene, and the structures of some of its derivatives, on the basis of such a structure. One difficulty is that double bonds between carbon atoms (four of which are in this structure) are known to undergo rapid chemical reaction with certain substances such as bromine. Benzene, however, does not react in this way.

One of Kekulé's great contributions was to suggest, in 1865, that in benzene the carbon atoms are at the corners of a regular hexagon. Today his structure is sometimes represented as follows:

```
        H
        |
  H     C     H
   \  //  \  /
     C     C
     |     ||
     C     C
   /  \\  /  \
  H     C     H
        |
        H.
```

This, however, was still not entirely satisfactory. For example, it fails to explain why benzene does not react rapidly with bromine and other substances. In 1874 Kekulé suggested that all the difficulties were avoided if one assumed that a "dynamic equilibrium" existed between two equivalent forms of the structure in which the double and single bonds are interchanged:

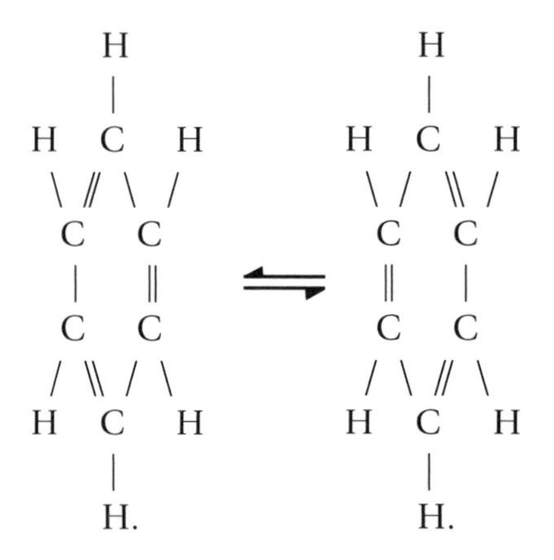

Kekulé's idea was that because of the equilibrium, with double bonds changing places with single bonds, none of the bonds would behave like double bonds. That idea, however, was still not entirely satisfactory.

A better way of looking at the benzene ring, due to a great extent to American chemist Linus Pauling (1901–94), is to think of it as existing in a hybrid, or *resonant*, state between the two forms, instead of rapidly changing from one to the other. One way of expressing this is to say that each of the carbon atoms has an order of 1.5. A single bond is said to have an order of 1, and a double bond of 2; the carbon-carbon bonds in benzene are halfway between single and double bonds.

For some time, little attention was paid to the actual shapes of molecules; they were simply represented as if they were flat. In 1874 an important contribution was made by Dutch chemist Jacobus Hendricus van't Hoff (1852–1911; see fig. 8). His main scientific work was in physical chemistry, but as a young man he made a contribution of great importance to organic chemistry. Some of his university studies were under Kekulé at the University of Bonn, and he later worked with French chemist Charles Adolph Wurtz (1817–84) at the Sorbonne in Paris. Before submitting his thesis to the University of Utrecht, and still not yet twenty-three, van't Hoff privately published a pamphlet on what has been called the *tetrahedral carbon atom*. His idea was that the four valence bonds issuing from a carbon atom do not lie in a plane, as in some of the previous structures shown in this chapter; instead, they point toward the corners of a

**Figure 8.** Jacobus Hendricus van't Hoff (1852–1911) first formulated general principles that explained the shapes of organic molecules. He also derived many thermodynamic relationships and applied them to practical chemical problems. (Edgar Fahs Smith Collection, University of Pennsylvania Library)

regular tetrahedron (see fig. 9). This idea of a tetrahedral carbon atom was not entirely original with van't Hoff; several chemists had considered the idea previously, but van't Hoff developed the idea in detail and on its basis explained many results that previously could not be understood.

A particularly important contribution made by van't Hoff in his pamphlet was to relate the three-dimensional structures of organic compounds to what is called their *optical activity*. An ordinary beam of light carries with it vibrations that occur in all directions at right angles to the path of the light (see fig. 10, *a*). It is possible, however, by passing ordinary light through certain materials, to produce what is called *plane-polarized light,* in which the vibrations are in one direction only (fig. 10, *b*). Earlier investigators had found that if plane-polarized light is passed through certain chemical compounds, the plane of polarization is rotated (fig. 10, *c*). Van't Hoff pointed out that, on the basis of his tetrahedral carbon atom, a molecule in which four different atoms or groups of atoms are connected to a carbon atom can exist in two different forms that are mirror images of one another; these are referred to as two different *configurations* of the molecule. This is shown in figure 11, in which the four groups are rep-

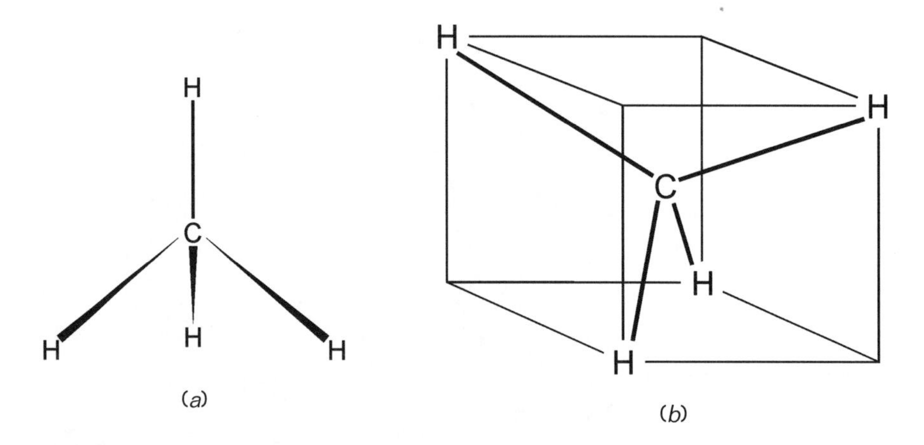


**Figure 9.** *a,* A representation of a simple carbon compound (methane, chemical symbol $CH_4$) showing the four bonds pointing to the corners of a regular tetrahedron; *b,* Another representation of the tetrahedral arrangement. The carbon atom is placed at the center of a cube, and the bonds point to four of its corners.

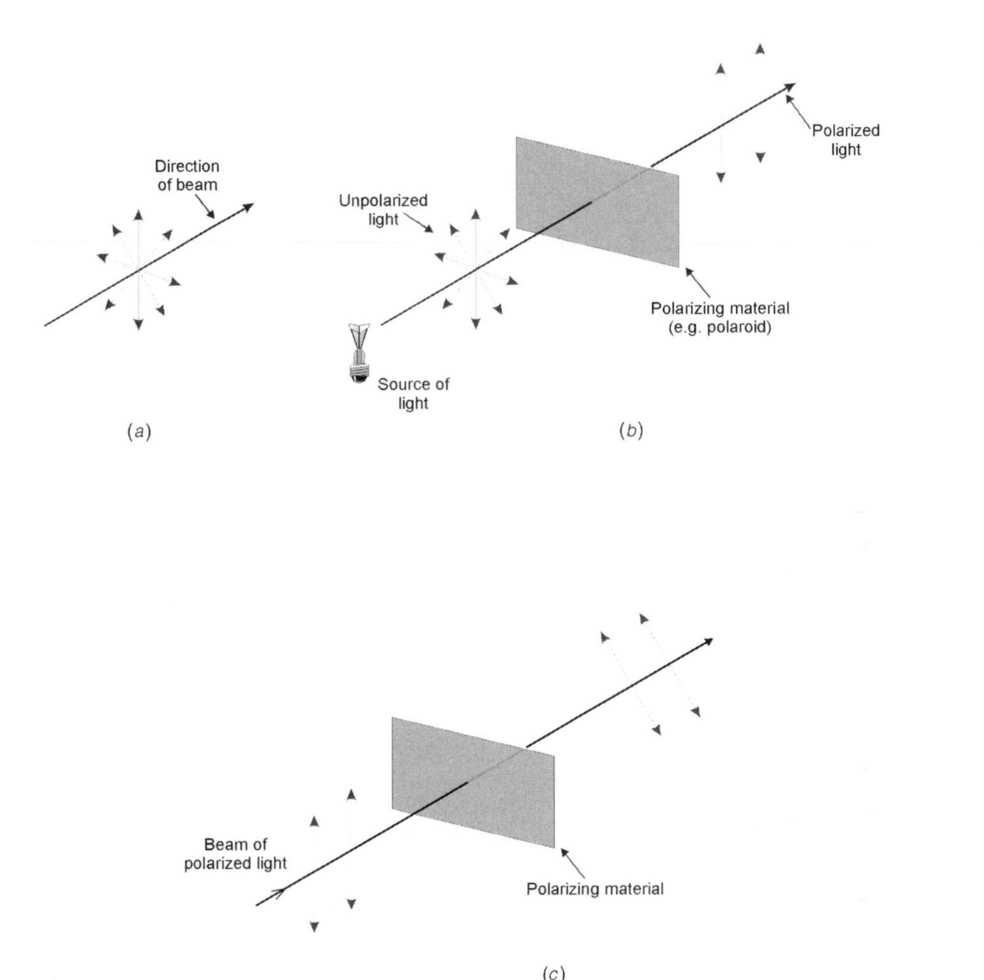


**Figure 10.** *a*, The vibrations that occur in ordinary light are at right angles to the direction of propagation of the light; *b*, the production of plane-polarized light by passing ordinary light through a polarizing material, causing the vibration to occur now in one direction only (up and down in this diagram); *c*, when plane-polarized light is passed through certain materials, said to be optically active, the plane of polarization is twisted, to the right or to the left.

resented as $R_1$, $R_2$, $R_3$, and $R_4$. It can easily be seen that if any two of the groups are identical to one another, it is no longer possible to have two different configurations. The important point made by van't Hoff was that whenever there are two different configurations that are mirror images of one another, there will be optical activity. That

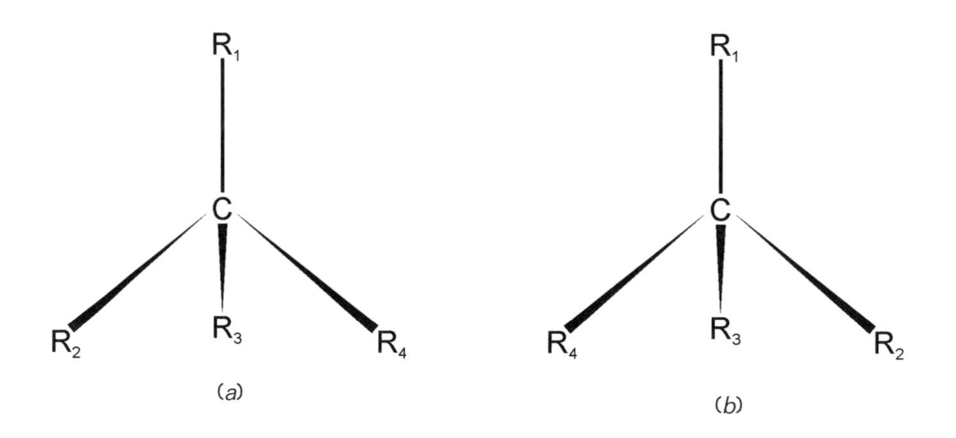


**Figure 11.** The two possible structures that can result from the attachment of four different atoms or groups of atoms to a carbon atom. The two structures cannot be superimposed on each other; each is the mirror image of the other. These two different forms are referred to as *configurations*. A carbon atom that is attached to four different groups is conveniently referred to as an *asymmetric carbon atom*. If any two of the attached groups are identical, there are no longer two distinct forms.

means that when polarized light passes through them, the plane of polarization is rotated. Optical activity thus arises when a carbon atom is attached to four different groups. Such a carbon atom is conveniently referred to as an *asymmetric carbon atom*. One form of the molecule will rotate the plane of polarization in one direction, while the mirror-image form will rotate it in the opposite direction.

Van't Hoff obtained strong confirmation of his ideas by applying them to some earlier work on a chemical called tartaric acid, done by eminent French chemist Louis Pasteur (1822–95). In tartaric acid there are two asymmetric carbon atoms, and van't Hoff showed that the different configurations that Pasteur had discovered were consistent with his theory.

Van't Hoff's PhD thesis, presented to the University of Utrecht in 1874, made no reference to the pamphlet he had published the year before on the tetrahedral carbon atom. It instead described a rather mundane piece of research in organic chemistry. The omission of the structural work seems at first sight surprising, but by the time van't Hoff submitted his thesis he had realized that his pamphlet was being violently attacked in some quarters. Scientists today find this rather remarkable, since van't Hoff was so obviously right in his proposals. Several prominent chemists agreed with him at once, but a particularly vitriolic attack on him was made by famous, but rather unpleasant, organic chemist Hermann Kolbe (1818–84). What is remarkable is that Kolbe did not attack van't Hoff on scientific grounds at all but by invective, referring to his ideas as "witchcraft." Apparently Kolbe thought it somehow improper even to think in terms of three-dimensional structures.

In spite of the hostility of Kolbe and a few other scientists, van't Hoff's ideas about the shapes of organic molecules were soon generally accepted. Chemists began to write three-dimensional structures for many of the simpler molecules containing carbon. With larger molecules, however, there were still difficulties. For example, it began to be recognized that molecules called proteins play an important role in living systems. Some of these were found to contain so many atoms that it was impossible, by ordinary chemical techniques, to determine even how the atoms were connected together; deducing their detailed three-dimensional structures from their chemical behavior was quite impossible. What was needed for this purpose was some new physical technique. Several such techniques were developed over the years; one that turned out in the end to be very fruitful involved the use of x-rays. Today the exact structures of a great many molecules, some of considerable complexity, are known in great detail.

So far we have represented valence bonds merely by straight lines and have said nothing about their nature. Not much could be done to explain the chemical bond satisfactorily until the electron had been discovered and after we had learned something about how electrons are arranged around atomic nuclei. In 1916 American chemist Gilbert Newton Lewis (1875–1946; see fig. 12) made a contribution of great significance: he suggested that the usual type of chemical bond involves a pair of electrons.

**Figure 12.** Gilbert Newton Lewis (1875–1946) made pioneering contributions to the theory of the chemical bond and also did significant work in thermodynamics as well as other fields. (Edgar Fahs Smith Collection, University of Pennsylvania Library)

Lewis was born in West Newton, Massachusetts, and was educated at the University of Nebraska and Harvard University, where he obtained his doctorate in 1897. He later worked in Germany with famous chemists Wilhelm Ostwald (1853–1932) and Walther Nernst (1864–1941). After a period at the Massachusetts Institute of Technology he went to the University of California in 1912, where he remained until 1946, when he died as a result of an accident in his laboratory: he was working with the highly poisonous chemical hydrogen cyanide, and some of it escaped into the room. He is remembered for making a great many important contributions to chemistry. He greatly clarified and expanded some of the fundamental principles of thermodynamics and developed a theory of acids and bases.

Lewis had first developed a rudimentary form of his theory of the chemical bond at Harvard in 1902 and had presented it to his students. In 1916, when he first published his theory, he was at the University of California at Berkeley. The important idea that Lewis introduced is that bonding can take place as the result of the sharing of electrons between two atoms. His original suggestion, later modified, was that groups of eight electrons, called *octets*, were stationed at the corners of cubes (see fig. 13, *a* and *b*). Bonding could occur by an overlap of the edges of two cubes (fig. 13, *c*). He recognized that in some molecules, both of the bonding electrons may originally have come from one of the atoms, an arrangement referred to later as a *coordinate link* or a *semipolar bond*. With suitable modifications, Lewis's model could also explain the tetrahedral arrangement of bonds emanating from a carbon atom (refer to fig. 9). The hydrogen molecule is a special case (see fig. 14). Each atom has one electron, and the molecule is formed by the pairing of these electrons. By residing between the protons, the two electrons reduce the repulsion between the protons, and a bond is possible.

In 1919 Lewis's ideas were taken up with great enthusiasm by American physical chemist Irving Langmuir (1881–1957), who

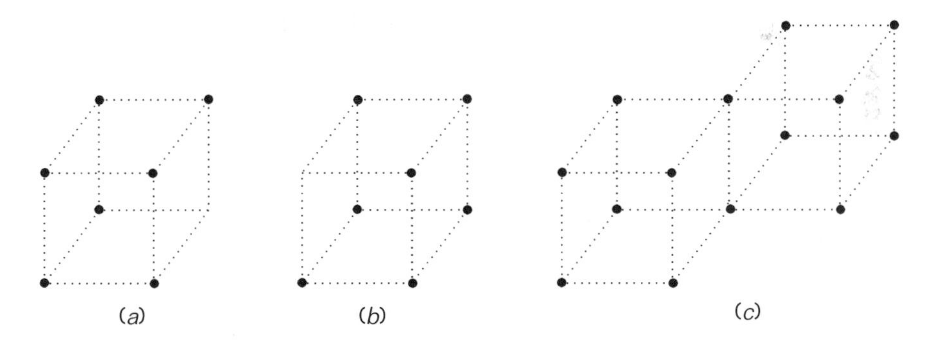


**Figure 13.** G. N. Lewis's original idea of chemical bonding, as he represented it in 1916. In some atoms the electrons arrange themselves at the corners of a cube, and he referred to them as *octets* of electrons. In *a* and *b* one electron is missing from the octet of each atom. In *c* the two atoms have come together to form a molecule; by coming together at the edge of the cube, each atom has completed its octet. Lewis's ideas were greatly modified and refined over the years and form the basis of modern ideas about the chemical bond.

Figure 14. Lewis's idea of the bonding in the hydrogen molecule, in which there are only two electrons, one originally from each atom. The molecule is formed by the pairing of these electrons. The two electrons, by residing between the protons, reduce the repulsion between the protons, making a bond possible. We know now that instead of residing between the two protons, the electrons simply have a higher probability of being between the protons.

already had a wide reputation, having done important work on solid and liquid surfaces that brought him the 1932 Nobel Prize for chemistry. Langmuir extended and to some extent modified Lewis's ideas, applying them to a wider range of substances. He also suggested a number of the terms that are used today, such as *covalence* and *electrovalence*. He published a dozen papers on the subject between 1919 and 1921, and he also presented the theory in numerous lectures he gave in the United States and elsewhere. Unlike Lewis, who was somewhat shy and diffident and not a good lecturer, Langmuir was always eloquent and compelling—to the extent that people sometimes complained that they found themselves convinced of things that they knew to be wrong! Consequently, although Langmuir always gave due credit to Lewis, the theory began to be known as the Lewis–Langmuir theory, which did not entirely please Lewis. In Britain, in fact, the theory was sometimes known as the Langmuir theory, which produced the following reaction from Lewis in a letter in 1926: "to persist, as they do in England, in speaking of the Langmuir theory of valence is inexcusable." Lewis was also resentful of the indifference of many of his immediate colleagues, particularly when he was developing his theory at Harvard from 1902 onward, and in 1929 he refused an honorary degree from Harvard.

Lewis was also somewhat disappointed by the attitude of the physicists, who regarded his theory as unreasonably crude. The physicists particularly objected to the fact that Lewis seemed to be regarding the electrons as fixed in certain positions and that the electron pair bond that he was proposing did not seem capable of binding the atoms together. In his turn, Lewis was critical of the physicists' idea of completely mobile electrons, which seemed incapable of explaining the undoubted fact, established by strong chemical evidence, that molecules have definite shapes. Over the years both Lewis's and the physicists' concepts became modified, and by 1923, when Lewis's famous book *Valence and the Structure of Atoms and Molecules* appeared, his ideas were entirely consistent with the ideas that the physicists themselves were then putting forward.[4]

The great importance of Lewis's theory is that it provided chemists with a valuable way of visualizing the electronic structures of atoms and molecules, and his ideas are still much used today. In spite of that, physicists have tended to ignore his contribution. A few physicists have appreciated what Lewis did, as illustrated by the following passage from *Wave Mechanics* (1945) by Walter Heitler (1904–81), who did important quantum-mechanical work on the covalent bond: "Long before wave mechanics was known Lewis put forward a semi-empirical theory according to which the covalent bond between atoms was effected by the formation of pairs of electrons shared by each pair of atoms. We see now that wave mechanics affords a full justification of this picture, and, moreover, gives a precise meaning to these electron pairs: they are pairs of electrons with antiparallel spin."

To many it has seemed surprising that Lewis did not receive a Nobel Prize in view of his great contributions to the theory of bonding as well as to thermodynamics. The fact that his bonding theory in its original form seemed naive to the physicists may have been the determining factor. In the 1920s and 1930s the nominations of Niels Bohr were given much consideration by the Nobel Prize committee, and many of his nominees received the award. He nominated numerous physicists whose work was of special interest to chemists, and Langmuir specifically for his work on surfaces, but he never nominated Lewis.

▲▲▲

The ideas developed about the atomic nature of matter that have been described in the present chapter from one of the great unifying principles in our harmonious universe. The whole of our universe, as far as we have been able to tell up to now, is made up of the same ninety or so elements, which remain the same in all space and in all time. We know this from spectroscopic studies of the outer regions of the universe, where we are seeing atoms that are not only far away but also belong to the distant past.

Atoms and the molecules that they form are largely held together by electrical—or, more precisely, *electromagnetic*—forces. The nature of these forces and how they affect the nature of our universe are discussed in the next chapter.

2

# THE ELECTROMAGNETIC UNIVERSE

It is not immediately obvious that light and electricity have much to do with one another. Humans were aware of the existence of light many centuries before they knew much about electricity. Electricity was not seriously investigated or indeed recognized at all until the eighteenth century—and even then was regarded for many decades as no more than a curiosity. Today, of course, most of us regard electricity as a necessity of life; we are highly dependent on it, and an extensive power blackout is seen as a disaster.

With the advance of science it has become evident that light and electricity are in reality closely related to one another and that magnetism forms the third member of the trinity. The elucidation of the connection between them took many years. Early in the nineteenth century the relationship between electricity and magnetism was discovered by Danish physicist Hans Christian Ørsted (1777–1851). Soon after that, Michael Faraday interpreted the connection between the two in terms of electric and magnetic fields. He discovered how electricity induces a magnetic as well as an electric field and how a magnet induces an electric as well as a magnetic field. At first his ideas were derided, but James Clerk Maxwell produced a brilliant synthesis of them and took them much further. He found that light can be understood in terms of an electromagnetic field, one that involves both an electric and a magnetic field. Even this is far from the whole story: it was later found that light sometimes behaves as if it were a beam of particles called *photons*.

To understand what has been discovered about light, electricity, and magnetism we need to go back a little further in time. The first important work on the nature of light was carried out in 1676 by eminent scientist Isaac Newton (1642–1727; see fig. 15). He was born in Lincolnshire, the son of a yeoman farmer who died shortly before Newton was born. Three years later his mother was remarried, to a clergyman whose Christian principles did not stretch to bringing up a stepson; instead, Newton was raised by his grandparents. This rejection as a child probably contributed to his paranoid behavior throughout his life: he was vain, manipulative, and humorless. Showing no inclination for a farming career he was sent to

**Figure 15.** Sir Isaac Newton (1642–1727), perhaps the greatest of all scientists, made fundamental contributions in many scientific fields. He also made important contributions to pure mathematics, discovering the binomial theorem and a system of calculus.

Trinity College, Cambridge, where he excelled in mathematics. In 1667 he was elected a Fellow of Trinity, and two years later became the Lucasian Professor at Cambridge. Newton carried out investigations on a wide range of problems, including the theory of equations, optics, the lunar and planetary orbits, gravitation, alchemy, chemistry, and theology—on which he spent more time than on anything else. His chemical experiments were carried out in a laboratory that he established in the garden behind his rooms in Trinity College. Newton's famous book, *Philosophiae Naturalis Principia Mathematica* (The Mathematical Principles of Natural Philosophy), was first published in 1687,[1] and his *Opticks* appeared in 1704.[2]

In 1696, after becoming afflicted with a severe nervous condition, Newton became warden of the Royal Mint, being promoted to master in 1699, with a substantial salary that enabled him to live in some luxury in London to the end of his life. He resigned his Cambridge chair in 1701, became president of the Royal Society in 1703, and was knighted in 1705. He is buried in Westminster Abbey.

Before Newton's time it had been noticed that colors are sometimes produced from white light, but there had been no careful investigation of the effect. In his rooms at Trinity College, Newton allowed a ray of sunlight to pass through a hole in the shutter of a dark room and then through a prism, where it was split into a band or spectrum of colored light (see fig. 16); a rainbow is a familiar example of such a spectrum, and it is sometimes said that there are seven colors in the spectrum, often identified as red, orange, yellow, green, blue, indigo, and violet. Newton also showed that the band of colors could be reconstituted into white light by passing it through another prism (also shown in fig. 16).

Newton's conclusion that white light is composed of a spectrum of colors was one of great importance. Artists had long been aware of the fact that any desired hue can be obtained by combining pigments in just three primary colors; for example, red, green, and blue pigments can be used as the primary colors. Important scientific work based on the idea of three primary colors was carried out in the early nineteenth century by Thomas Young (1773–1829; see fig. 17), who was born in Somerset. He studied medicine at the universities of

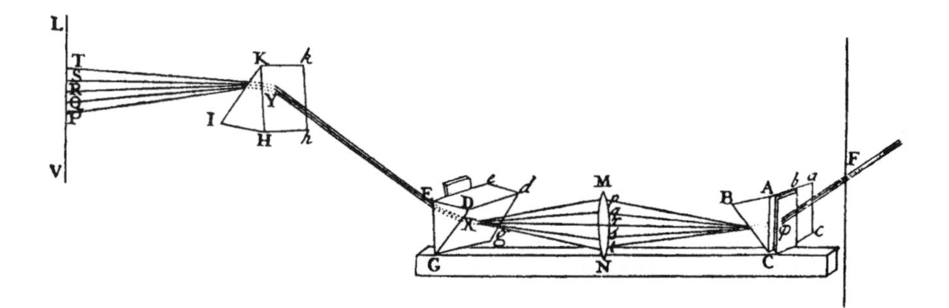

**Figure 16.** The illustration of Newton's spectrum experiment that appeared in his treatise *Opticks* (1704). Sunlight enters the room through a hole, *F*, in the window shutter. The first prism produces the spectrum, *pqrst*. The lens, *MN*, focuses it to enter a second prism, where the rays are reconstituted into white light, *EY*. A third prism produces a spectrum again.

London, Edinburgh, and Göttingen and then practiced medicine and wrote medical treatises. Young was remarkably talented, and his interests were very broad. Because he never quite finished anything that he began, he did not accomplish as much as he should have. Toward the end of his life he said that he had never spent an idle day in his life but admitted that his main contributions had been to make "acute suggestions" and to leave it to others to work out the details.

Young was an infant prodigy, who could read at the age of two, studied Latin at six, and while still in his teens had a working knowledge of more than a dozen languages, including Hebrew, Chaldean, Arabic, and Turkish. In 1801 he became the first professor of natural philosophy at the Royal Institution but was not a successful lecturer. For this reason, and because his relations with the Institution's managers became strained, he resigned in 1803. His main scientific contributions were to clarify the understanding of human vision, to discover important properties of light, and to outline a wave theory of light. Before developing his wave theory, however, he returned to his long-standing interest in languages and hieroglyphics.

One of Young's contributions to physiological optics was to show that the lens of the eye changes its focal length and so assists in the eye's ability to *accommodate*, that is, to focus on objects at

**Figure 17.** Thomas Young (1773–1829) made contributions of great importance to physiological optics and to the establishment of the wave theory of light. (Edgar Fahs Smith Collection, University of Pennsylvania Library)

various distances. Previously it had been thought that accommodation was due solely to changes in the curvature of the cornea (which in fact do not occur) and to changes in the depth of the eyeball (which do occur). Young then went on to study color vision. He concluded that the eye contains three color receptors, corresponding to red, yellow, and blue, and that the eye recognizes colors by the superimposition of images from these receptors. When lights of these three colors are mixed together in suitable proportions, the result is white light.

Young also made contributions of great importance to our understanding of the nature of light. Newton had interpreted the result of his experiments for the most part in terms of the theory that light consists of a stream of particles, which he called corpuscles. In 1665 Newton's archenemy, Robert Hooke (1635–1703), had put forward a wave theory of light, but he did not develop the idea to any extent. According to Hooke, the vibrations occurring in a light beam are in the same direction as the light path, whereas the modern theory is that they occur at right angles to the light path (refer to fig. 10, *a*, in the previous chapter). A little later, eminent Dutch investigator Christiaan Huygens (1629–95) brilliantly developed Hooke's wave theory and showed how it could explain the reflection and refraction of light.

Newton gave careful consideration to both the corpuscular and wave theories and did not reject either of them. Indeed, he tended to favor ideas that were partly wave and partly corpuscular, which is rather remarkable, since that is the modern view. Newton's main difficulty with the wave theory is that he could not see how it could explain the fact that light travels in straight lines; he thought that when light encountered an obstruction, the waves would be sent out in different directions. There is indeed some spreading of this kind, but because the waves are so small, the spreading is so slight as to be difficult to detect. Newton changed his views from time to time, but on the whole he favored the corpuscular theory. Because of his great prestige, this was the theory generally accepted until Young's work at the beginning of the nineteenth century.

Young's significant contribution began with his discovery of the *interference of light*. He passed a beam of white light through two slits in a screen, with another screen placed beyond the first (see fig. 18). Where the rays overlapped on the second screen, he saw a series of colored bands. If instead of white light he used light of one color (called monochromatic light), the bands were instead alternately light and dark. He explained the bands obtained with light of one color in terms of the superimposition of the waves from the two slits. If the distances traveled by the two waves is the same, the crests of the waves will be superimposed, and the intensity of the light will be doubled. If, on the other hand, one wave has half a wavelength farther to travel than the other, the crest of one wave will coincide with the trough of the other, and darkness will result; this is what he

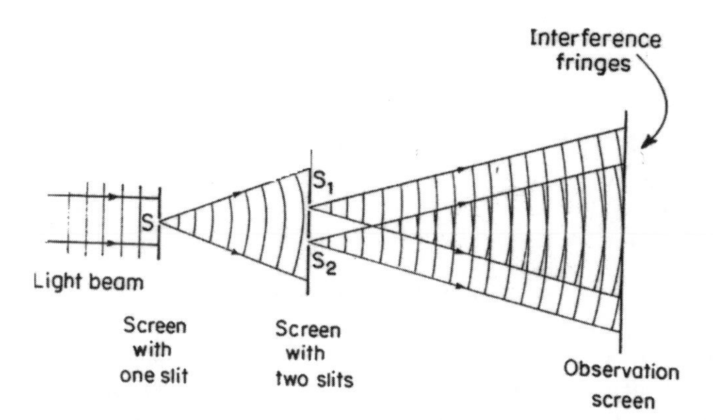


**Figure 18.** A schematic representation of Thomas Young's interference experiment. The light waves spreading from slit *S* reach the two slits $S_1$ and $S_2$. The light waves coming from these two slits give rise to interference fringes on the observation screen. If the light is monochromatic (i.e., involving only a narrow range of wavelengths) the observation screen shows light and dark bands; otherwise, colored bands are seen.

meant by interference. Figure 19 shows the shape of a light wave and shows how two waves can annihilate one another. When white light is used, the different colors interfere at different positions, and the result is that the bands are colored. This work was of great importance, since it provided strong evidence for the wave theory of light.

The wave theory leads to the result that the wavelength multiplied by the frequency (the number of complete vibrations that occur in a unit of time) is equal to the speed with which the light travels:

$$\text{wavelength} \times \text{frequency} = \text{speed of light}.$$

From the dimensions of his apparatus and the breadth of the bands, Young was able to calculate the wavelengths of the different-colored rays. He found that the wavelengths are exceedingly small compared with the lengths of ordinary obstacles in the path of a beam of light. The light at one end of the visible spectrum is red, and Young found that red light has a wavelength of about 700 billionths of a meter ($7 \times 10^{-7}$ m, or 700 nanometers, or 7,000 angstroms). Violet radiation is at the other end of the visible spectrum, and Young found it to

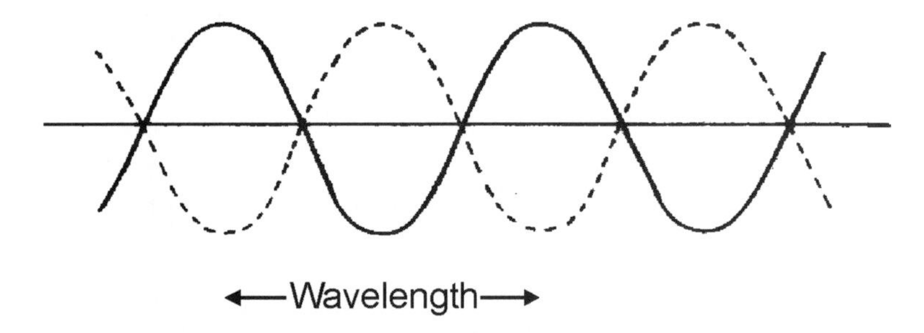


**Figure 19.** A representation of wave motion. The wavelength of the light is the distance between neighboring crests or troughs. The dashed curve shows another wave such that its troughs coincide with the crests of the first wave. The two waves would annihilate one another. The frequency is the reciprocal of the time taken for the complete vibration to occur. The product of the frequency and the wavelength is the speed of light.

have a wavelength of about 400 nanometers (4,000 angstroms), or roughly half that of red radiation.

The fact that the wavelengths are so small compared with ordinary objects overcomes a difficulty that Newton had found with the wave theory of light. He thought that if a wave arrived at the edge of an object it would be diffracted and would not continue along the same straight line. Because of the shortness of the wavelengths, however, the extent of the diffraction is slight, such that the diffraction of light can only be detected under specially contrived conditions.

Young did not pursue the problem much further, since he had so many other interests. It was left to French physicist Augustin-Jean Fresnel (1788–1827) to develop the wave theory mathematically to a high degree of perfection. One of the most important changes he made was to regard the vibrations as transverse (at right angles) to the direction of the rays. Robert Hooke's idea that the vibrations are in the direction of the rays presented a difficulty with the experiments on the polarization of light (refer to fig. 10, *c*). Rotation of the plane of polarization of light can only be explained if the vibrations are at right angles to the beam.

The measurement of the speed of light is a matter of great importance. Aside from allowing the frequency to be calculated from the wavelength, the speed allows us to distinguish between the corpuscular theory and the wave theory favored by Young. If the corpuscular theory applied, the speed would be greater in a medium such as water than it is in air. According to the wave theory, however, the speed would be less in water. It is therefore important to determine the speed in water as well as in air. One of the first estimates of the speed of light in air was made in 1675 by Danish astronomer Olaus Rømer (1644–1710). He obtained his value by observing the satellites of the planet Jupiter. The time of their revolution about the planet was known accurately, and the distances of the planet from Earth were known at various times. Rømer realized that if light did not travel instantaneously but at a finite rate, the eclipses of the satellites should appear to take place earlier when the planet was nearer to Earth than when it was farther away. From the expected and observed times he calculated a value for the speed of light, but it was too low by about 40 percent. The essential thing he demonstrated is that light does not travel with infinite speed, which until then had been considered possible.

Reliable measurements of the speed of light were first made by two remarkable French physicists, Léon Foucault (1819–68) and Hippolyte Fizeau (1819–96). They were both born in Paris, Foucault just four days before Fizeau. In 1884 they began to work together on the speed of light, devising a satisfactory experimental system, but a few years later dissolved their partnership over a personal dispute. They continued their work independently and met with success almost simultaneously in 1849. Both men paid special attention to determining whether light travels more rapidly in air or in water, since this provided a way of deciding between the corpuscular and wave theories.

In determining the speed of light in air, Fizeau was successful slightly before Foucault. He caused a beam of light to travel the double distance between his father's house in Suresnes, a suburb of Paris, and the top of a hill in the Paris suburb of Montmartre, about 8 km (5 mi) away. The speed was measured by ingenious instruments involving rapidly rotating toothed wheels. Fizeau's first value was 315,000 km per second, which is rather high compared with the modern value of 299,792 km (186,291 mi) per second. The experimental method, however, was reliable, and was used later in more precise determinations.

Foucault and Fizeau also compared the speed of light in air and water using quite a similar apparatus, and in this Foucault was the first to be successful. On April 30, 1850, Foucault announced his discovery that light travels slightly faster in air than in water, and a few days later the same result was obtained by Fizeau. This result favors the wave theory over the corpuscular theory.

Another important development in the understanding of light was made by James Clerk Maxwell (1831–79; see fig. 20), one of the greatest scientists of all time. He was born in Edinburgh to prosperous parents, his father being laird of Glenlair, a position that James inherited on his father's death. Much of James's boyhood was spent at the manor house at Glenlair, which is near Corsack in County Galloway, in the southwest of Scotland. At the age of ten he

**Figure 20.** A photograph of James Clerk Maxwell (1831–79) taken during his tenure of the Cavendish Professorship at Cambridge. He made pioneering contributions in several fields and is particularly remembered for his theory of electromagnetic radiation. He is also noted for his treatment of the distribution of molecular speeds in a gas, work that led to the development of important statistical methods in science.

was sent to Edinburgh Academy where, on account of his broad Scottish accent, which he retained throughout his life, and his homespun ways, he was somewhat ridiculed by his fellow schoolboys, who inappropriately nicknamed him "Daftie." In 1847, at the age of sixteen, he became a student at the University of Edinburgh. During his undergraduate years he did a good deal of research, particularly on color theory, and he also began to think about the implications of Faraday's great work on electricity and magnetism.

After taking his Edinburgh degree in 1850 Maxwell became an undergraduate at Cambridge, first at Peterhouse and later at Trinity College. While still an undergraduate at Cambridge he continued his work in color theory. In 1855 he was elected a Fellow of Trinity, and from 1856 to 1860 he was professor of natural philosophy (i.e., physics) at Marischal (pronounced "Marshal") College, Aberdeen. It was here that he did the bulk of his work on color theory; it was essentially completed during the next five years, from 1860 to 1865, when he held his second appointment at King's College, London. To some extent Maxwell's work on color involved extending the work of others, so it was not as highly original as some of the rest of his work. At the same time, there is no doubt that Maxwell was one of the great pioneers of the theory of color and of color physiology, being the first to establish a quantitative basis for the subjects.

His investigations showed that although red, yellow, and blue make an acceptable set of primary colors, as proposed by Thomas Young, a more appropriate set is provided by red, green, and blue. Maxwell distinguished clearly, for the first time, between *hue* (or color), *tint* (or saturation), and *shade* (lightness or intensity). The hue, which is defined by the wavelength of the light, simply means what we usually call the color; the hue may be red, yellow-green, or blue, for example. Tint, or saturation, has to do with the extent with which the hue has been influenced by the addition of other hues. If, for example, blue is the only hue present, the saturation is said to be 100 percent; the tint is pure blue. The shade, or intensity, is related to the amount of light that reaches the eye. If an object is illuminated by white light and it reflects all the incident light, it has maximum brightness and appears white. If it reflects no light, it appears black. In the case of a self-luminous object, the shade can be called the brightness.

For his investigations on color, Maxwell used a number of observers, including his wife, and some of them suffered from color blindness. All of his results convinced him of the essential correctness of Young's three-receptor theory. Some of his investigations were carried out using color boxes of his own design. While Maxwell was professor at King's College, London, from 1860 to 1865, he worked on one of these color boxes in the attic of his house in Kensington. When his neighbors observed him constantly staring into this box, which was painted black and looked like a coffin, he gained quite a reputation as an eccentric.

On May 17, 1861, Maxwell made an important demonstration at the Royal Institution, in the presence of Michael Faraday, among many others. He was able to project on a screen an image of a colored object, a ribbon made into a rosette with red, green, and blue stripes. Three black-and-white photographs were taken, through red, green, and blue filters, and negatives were prepared from them. Three projectors, lighted by red, green, and blue lamps, were then used to produce superimposed colored images from the corresponding three black-and-white positives. The result was a reasonably good color image of the ribbon. Maxwell is sometimes said to have produced the first colored photograph, but it seems better to call it the first *photoimage*, as no permanent record had been obtained. The work was, however, critical in leading the way to color photography. An interesting development of the idea was made much more recently by American scientist and inventor Edwin Land (1909–91). It is well known that our perception of a color is strongly affected by neighboring colors. A colored photograph or painting may therefore be completely acceptable even though the colors of individual small regions of it may not exactly correspond with the original. Land developed a theory, known as his *retinex theory*, that took these perceptions into account. His practical conclusion from his studies was that one could simplify color photography by going from the three-color system to a two-color system that in practice gave equally satisfactory results. He was then able to produce in 1963 a color film that could be developed inside the camera (the Polaroid Land Camera) within minutes.

▲▲▲

The existence of what came to be called electricity was known to the ancient Greeks. Statesman and scientific investigator Thales of Miletus (c. 636–c. 546 BCE) made many important contributions to knowledge. He discovered that amber, when rubbed, acquires the property of attracting small pieces of pith or cork. Many centuries later William Gilbert (1544–1603) extended these observations, showing that other substances, such as glass and sulfur, show the same attracting properties when rubbed. To explain this behavior he coined the word *electric* from the Greek ἤλεκτρον (*ēlektron*), meaning amber. Gilbert, who became physician-in-ordinary to Queen Elizabeth and to King James I, also made pioneering investigations on magnetism.

Early work on electricity was concerned with static electricity. Machines were invented for generating electric charge, and electricity was stored in what were called *Leyden jars*. These were the prototypes of the modern *capacitors*, which are two pieces of metal close to one another but separated by an insulator, which does not allow electric charge to pass. If a capacitor is highly charged with electricity, wires connected to the two plates will produce a spark if they are brought together; a person holding the wires may receive a dangerous electric shock. Capacitors are widely used in research and technology today.

At first, electricity was regarded as little more than a curious and unimportant phenomenon. In 1729 Stephen Gray (1666–1736) distinguished between conductors and nonconductors (insulators) of electricity. Soon afterward a two-fluid theory of electricity was formulated. Bodies that were not electrified were supposed to contain equal quantities of "vitreous and resinous fluids." An electrified body was supposed to have gained an additional quantity of one fluid and to have lost an equal amount of the other, the total amount of electrical fluid remaining the same. In those times the idea of an imponderable fluid was a popular one; heat, light, magnetism, and electricity were all commonly regarded as invisible and weightless fluids that could pass with ease from one body to another.

Important advances in the understanding of electricity were made by eminent American scientist and statesman Benjamin Franklin, whom we already met in chapter 1. He predicted, and confirmed experimentally in 1747, that if a metal rod is connected to the earth (i.e., "earthed," or "grounded"), and is pointed toward a body that is highly charged with electricity, the body becomes discharged, losing its electricity. This observation led to Franklin's important invention of the *lightning rod,* one of which he installed on his own home in Philadelphia in 1749. In 1752 Franklin carried out his famous experiment, in which he flew a kite bearing a lightning rod connected electrically to the ground.[3] Several people who repeated the experiment were killed, but Franklin was lucky (and careful!). His experiment showed that thunderclouds result from the accumulation of electricity. In that same year he attached a bell to the lightning rod on his house to let him know of any electrification from passing clouds.

The lightning rod, which was never patented by Franklin, soon attracted wide attention. After a London church was struck by lightning, lightning rods were installed on St. Paul's Cathedral in 1769, but they inspired some controversy. Some argued that their installation acted against the work of God: if an angry God wished to destroy a building, so be it. There was also controversy about whether lightning rods should be pointed, as recommended by Franklin, or rounded. King George III allowed himself to be persuaded by Tory politicians that the pointed rods were a plot by Franklin and other British colonists in America to destroy government buildings in England and that rounded rods were preferable. Although a committee of the Royal Society considered the matter and also recommended pointed rods, His Majesty ordered rounded rods on government buildings. There is a story that Sir John Pringle, president of the Society, commented that "the laws of nature are not changeable at Royal pleasure" and that the king then angrily pressed him to resign.

Franklin's investigations led him to reject the two-fluid theory in favor of a one-fluid theory of electricity, according to which there is a single electrical fluid that nonelectrified bodies were considered to possess in a certain normal amount. A body having more than this amount was said by Franklin to be positively charged, while one deficient in the fluid was negatively charged. This turned out to be a somewhat unfortunate choice: a body having an excess of electrons is in fact negatively charged, and what was later described as a flow of current through a wire in one direction is a flow of electrons in the opposite direction. Many of Franklin's conclusions about electricity were set forth in his book *Observations on Electricity, Made at Philadelphia in America,* which was published in London in 1751.

An important contribution to electricity research was made by French military engineer Charles Augustin de Coulomb (1736–1806), whose early work had been on the construction of military fortifications. In 1784 he constructed a torsion balance, consisting of a long bar hung at its center from a wire. With it he showed that the force of electrical attraction is proportional to the product of the electrical charges and inversely proportional to the square of the distance between them. He also showed that the same law applies to magnetic poles. This law is now known as *Coulomb's law.*

The first work that led to investigations of electric currents was done in the late eighteenth century by Luigi Aloisio Galvani (1737–98), who described his findings in a book published in 1791. Galvani found that when metals are inserted into frogs' legs, muscular contractions occurred, and he attributed the effect to "animal electricity." Others had made similar observations, and Galvani's interpretations of his results were incorrect, but his name has been preserved in a number of expressions. A flow of electricity is sometimes called galvanic electricity, and a battery or cell producing electricity is often called a galvanic cell. Iron that has been treated electrically so that it is protected by a layer of zinc is called galvanized iron, and the expression is used even when a similar type of protection is achieved without the use of electricity. An instrument used for detecting and measuring current is called a galvanometer, and a person may be galvanized into action, not usually by application of electricity.

More important contributions were made by Italian physicist Alessandro Volta (1745–1827) of the University of Padua. His great achievement was to show that the production of electricity does not necessarily involve living systems. He recognized that in Galvani's experiments the frog's leg was not essential to the production of electricity but that it merely acted as a detector of electricity; he realized that the electricity was being produced from the metals and solutions present. In 1800 Volta invented what is called a "Voltaic pile," consisting of a series of discs of two different metals, such as silver and zinc, separated by paper moistened with brine (see fig. 21). He found that a current of electricity was produced, and the same year his discovery was published in the *Philosophical Transactions of the Royal Society*.[4] Volta explained the electricity produced by his piles simply in terms of contact between the metals and the solution, but this explanation is inadequate. Chemical reaction is occur-

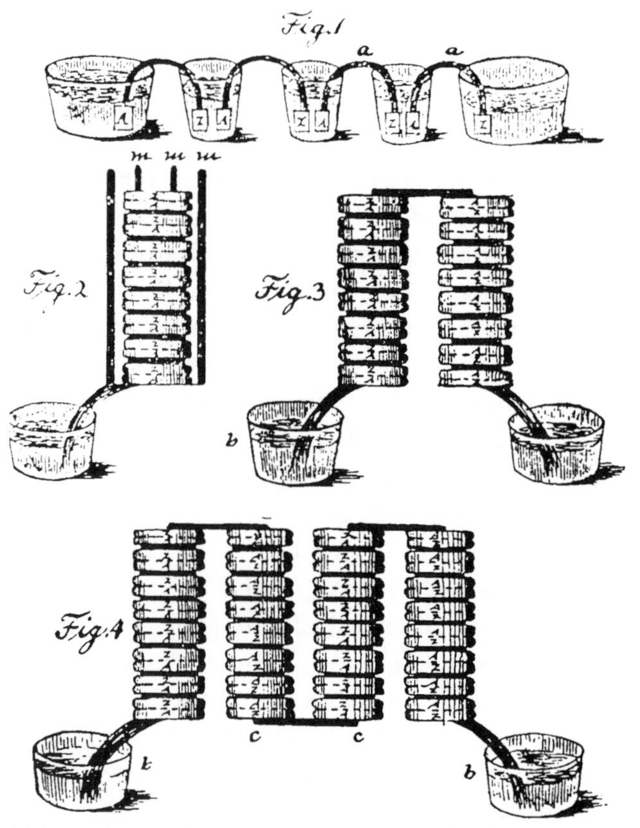


**Figure 21.** Voltaic piles, as illustrated in Alessandro Volta's first paper on the subject. For his contributions, the emperor Napoleon made Volta a count in 1801.

ring in the Voltaic pile, and the source of the electrical energy is the chemical energy released as the chemical processes occur.

At first the relationship between static electricity and what was produced in a Voltaic pile was not at all clear. It was only after more work had been done with electric batteries that it was realized that they produced a flow of the same kind of electricity that could be stored in Leyden jars. Only very much later was the mechanism of the flow understood—for example, that the flow of electricity along a wire is a flow of electrons, the existence of which was not recognized for decades.

Even before Volta's letter appeared in print, another paper was published that described experiments that showed that the electricity generated in a Voltaic pile was capable of bringing about the production of hydrogen and oxygen gases from water.[5] The author of this paper was William Nicholson (1753–1815), the founder and editor of the *Journal of Natural Philosophy, Chemistry, and the Arts*; it was usually known as "Nicholson's Journal." The experiments had been done in collaboration with his friend Anthony Carlisle (1768–1840), a fashionable London surgeon who was later knighted and became president of the College of Surgeons. The two men had constructed Voltaic piles of their own, often using half-crowns (a common silver coin in Britain, still remembered by older people) as the silver discs; one of their piles, for example, consisted of "17 halfcrowns, with a like number of pieces of zinc, and of pasteboard, soaked in salt water." They then inserted wires from the two ends of their pile into a dish of water containing a little acid or other substance. With platinum or gold wires they found that hydrogen gas was liberated at one wire and oxygen gas at the other. This process became known as *electrolysis*.

Nicholson and Carlisle's discovery that electricity can cause water to produce hydrogen and oxygen created as great a stir as any scientific discovery ever made. The surprise was not so much at the fact that the gases were produced, but *where* they were produced. It seemed that if the gases were produced at all, they should be produced in one place. We can look at the matter as follows: imagine the water at the wire at which hydrogen is evolved (i.e., released). Why is not the oxygen, also presumably formed from the decomposition of water, evolved at the same place? Why and how does it apparently "burrow" its way through the solution and appear only at the other wire? The dilemma could not be solved satisfactorily for many years, until evidence was obtained that charged species, such as hydrogen ions $H^+$ and hydroxide ions $OH^-$, exist in solution. The electrolysis of water could then be explained in terms of hydrogen ions moving toward one electrode and hydroxide ions toward the other. Later, Michael Faraday did much work on electrolysis, to be discussed later.

The work of Volta and of Nicholson and Carlisle stimulated many further investigations. An interesting electric pile was designed by George John Singer and described in his book *Elements of Electricity and Electro-Chemistry*, which was published in London in 1814. The pile consists of a large number of pairs of silver and zinc discs, separated from one another by manganese dioxide. What is remarkable about this pile is that one built in 1840, probably to the same specifications, is still in operation over 1½ centuries later. It was installed at Oxford by Robert Walker, a distinguished teacher of science and a textbook writer. The pile is to be seen in the Clarendon Laboratory at Oxford, and it continues to ring a bell to this day. It has to be explained to those who go to see it that there is no perpetual motion; the bell is bound to stop eventually, and it seems possible that the clapper will wear out before the electrochemical energy runs out. It is estimated that the voltage is about two thousandth of a volt and the current about 1 billionth ($10^{-9}$) of an ampere (or 1 nanoampere). Although the pile is referred to as a dry pile, it cannot be really dry, as then it would not work. One of the reasons for its longevity is that the instrument makers sealed it so well that the tiny amount of moisture present cannot escape.

Voltaic piles were awkward to use; it was more convenient to immerse rods or plates of two different metals in a bath of acid or brine (salt solution). Such a device is usually called a *battery*, although strictly speaking it should be called a *cell*; the word battery should be reserved for several cells connected together, although today this convention is often violated. In the hands of James Prescott Joule electric cells played a vital role in leading to the first law of thermodynamics. They were also important in leading to the discovery of some of the chemical elements. They were employed, for example, at the Royal Institution by Humphry Davy (1778–1829) to isolate several elements. On October 6, 1807, he electrolyzed fused potash and obtained the element potassium, and a week later he electrolyzed fused soda and produced pure sodium. In the following year he isolated three more elements; calcium, strontium, and barium; he also obtained magnesium but only in an impure form. On the basis of work of this kind, Davy advanced the hypothesis that electrical attractions are responsible for the formation of chemical compounds, which turned out to be correct, although the situation is much more complicated than he thought.

In 1808, Davy brought together two carbon rods connected to the poles of a battery and produced a brilliant arc. This can be said to be the first lamp powered by electricity, but it was not for another half-century that arc lamps were used for public lighting. In the same year Davy launched an appeal for funds that would enable the Royal Institution to construct a battery of a large number of cells, capable of producing what has been estimated to be five thousand volts. (This unit, named in honor of Volta, was not introduced until later.)

Several useful electric cells were devised by John Frederick Daniell (1790–1845), professor of chemistry at King's College, London. One of Daniell's cells, constructed in 1836, consisted of an outer vessel containing copper sulfate solution, inside which was an ox-gullet bag containing a rod of amalgamated zinc in dilute sulfuric acid; it produced about 1.1 volts. In a later modification the ox-gullet was replaced by a porous pot that contained the acid and the zinc electrode. In 1839 Daniell put together a battery of seventy of his cells and produced a brilliant electric arc, the light from which caused some skin blistering as well as eye injuries to himself and others.

William Robert Grove (1811–96) also devised some useful electric cells. In 1838, while a professor at the London Institution, he devised a cell consisting of zinc in sulfuric acid and platinum in nitric acid, a cell that produced 1.8 to 2.0 volts—greater than that given by Daniell's cells. Cells of this type began to be used by Michael Faraday in his demonstrations at the Royal Institution. First described in 1839, Grove also devised what he called a *gas voltaic battery*, consisting of platinum electrodes immersed in acid solution, with hydrogen bubbled over one electrode and oxygen over the other. The source of the voltage is the energy released by the combination of the two gases. Later Grove used other combinations of gases, such as hydrogen and chlorine as well as oxygen and carbon monoxide. Cells of this type are today called *fuel cells*, and they are still the subject of much research to make them more effective for commercial use.

Grove, who later practiced as a lawyer and became a judge, was a rather colorful individual who was fond of undertaking spectacular and sometimes dangerous experiments. At a lecture he gave to the Royal Society in May 1843 he connected together fifty of his fuel cells and caused the current to pass through five people holding hands, giving them all a shock. When he shorted the battery of cells, using charcoal points, a brilliant spark was produced. He gave a total of fourteen "Discourses" at the Royal Institution, and at one of them, on March 13, 1840, he managed to destroy Michael Faraday's pocketknife. Using a battery of fifty cells, each having large metal plates, Grove produced an arc about 3 cm long. Faraday lent him his knife to aid in the demonstration, and we are told that the "large blade was instantly deflagrated." We are not told what Faraday said about the incident; in view of his deeply religious nature we can assume that his comments had no need of expletive-deletion.

Later in the century came the introduction of dry cells, the development of which sprang from the work of French engineer Georges Leclanché (1839–82). He did not construct a dry cell himself, but in 1868 he devised a cell in which rods of zinc and carbon were dipped into a solution of ammonium chloride. Wet cells of this type were used well into the twentieth century for doorbells and other purposes, where electricity is required only occasionally and for short periods. Some of the familiar dry cells now used very widely are a development of the Leclanché cell.

The cells mentioned so far are called *primary cells*. They cannot be regenerated by electrical means but only by replacing the solutions or *electrodes* (as the "poles" were later called by Faraday). Later, storage cells or accumulators were developed, which could be recharged by the passage of a current. Today they are used widely in automobiles and will find an even greater application when hybrid electric-gasoline cars become more common.

▲▲▲

In 1820 Hans Christian Ørsted (1777–1851; see fig. 22, *a*), professor of physics at the University of Copenhagen, brought a compass needle close to a wire through which an electric current was passing. He found that the needle turned in an unexpected way. The poles of the magnet were neither attracted nor repelled by the current, but rather moved in a direction at right angles to the expected direction (fig. 22, *b*). When the direction of the current was reversed, the needle turned sideways in the opposite direction.

This famous discovery is remarkable for having been made during a lecture to a class of students. Ørsted and his assistant had set up the experiment but had not had time to try it out before the students arrived. Ørsted first decided to defer the experiment until later, but during the lecture he began to feel confident that it would

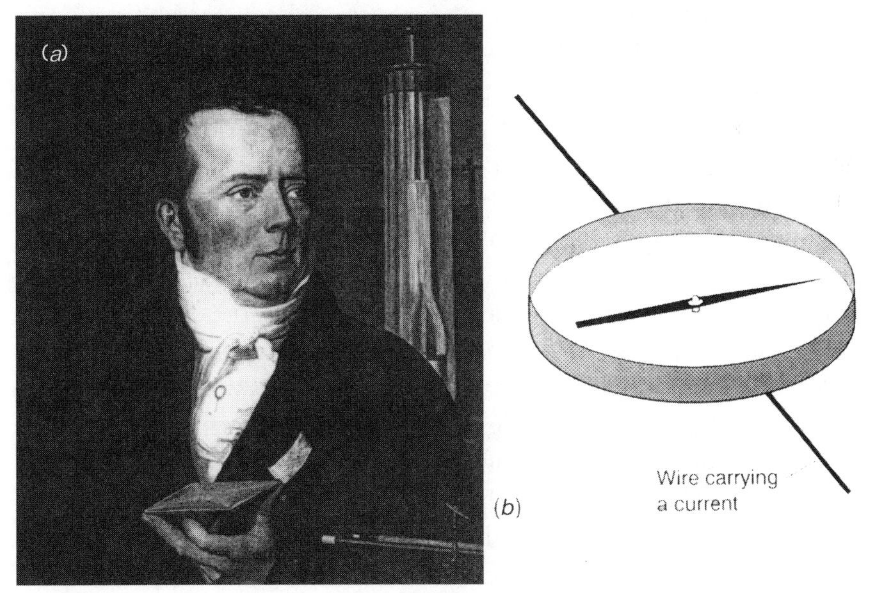


**Figure 22.** *a*, Hans Christian Ørsted (1777–1851), the discoverer of electromagnetism; *b*, a representation of Ørsted's experiment. The magnet tends to turn toward a position at right angles to the wire carrying a current.

work and performed it successfully. Another remarkable feature is that, since the deflection of the needle had been small and hardly visible to the audience, Ørsted was so little impressed by the result that he performed no further experiments on the subject for three months. Then, having confirmed the effect, he sent a four-page account of it to many leading scientific journals and to several other scientists. To make it more widely known he wrote his announcement in Latin, and it appeared, translated into various languages, in a considerable number of journals. Further publicity was given to the discovery by the fact that the distinguished scientist and statesman François Arago (1786–1853) called attention to the discovery at a meeting of the Académie des Sciences in Paris on September 4, 1820. At first Arago was skeptical about the result, but a week later he repeated Ørsted's experiment and became convinced. This experiment, showing for the first time a connection between electricity and magnetism, can be called the birth of *electromagnetism*.

At the time, in the scientific tradition established by Newton in particular, theories were formulated in terms of forces acting in straight lines between points; these were called *central forces,* and they were assumed to follow what is called the *inverse square law,* the force being inversely proportional to the square of the distance between the points. The fact that the magnetized needle was subjected to a deflecting force (i.e., it moved in a sideways direction) was therefore surprising. It suggested a force not acting in a straight line but circularly, which most scientists thought to be unreasonable. However, within a short time many investigators had confirmed the experimental result obtained by Ørsted.

Within only a few days, much progress, along both experimental and theoretical lines, was made by French physicist André-Marie Ampère (1775–1836; see fig. 23, *a*). Ampère was a prodigious worker; his life was a tragic one, and he deliberately drowned his sorrows in unremitting labor. At the age of eighteen he was forced to witness the guillotining of his father, an experience that left him highly traumatized; for a long time he wandered listlessly. A happy marriage in 1799 later raised his spirits, but the death of his wife four years later was a blow from which he never recovered. A second marriage to a woman who, with the connivance of her relatives, made his life unbearable, ended in divorce. Even his children by his first wife were a great trial to him, and he had constant financial problems.

Ampère was held in high regard by his colleagues and was elected a member of the exclusive Institut de France. He was an eccentric and absentminded man, which resulted in many stories about him. At one of the meetings of the Institut he failed to recognize the emperor Napoleon, who good-humoredly did not take offense. Instead he said that he knew a sure way of making himself recognizable and asked Ampère to lunch on the following day; Ampère unfortunately forgot to appear. Ampère kept a cat and had a hole of suitable size cut in a door so that it could go in and out. When he (and perhaps the cat) later acquired a kitten, he had a smaller hole cut in the door, not realizing until it was pointed out to him that the second hole was unnecessary. (This story, it must be admitted, has also been told about Newton and Bunsen; quite possibly it happened to all of them.)

Ampère was in the audience at the Acadèmie des Sciences in Paris on September 4, 1820, when Arago announced Ørsted's discovery. Ampère had an exceedingly agile mind, and within a week he had made careful quantitative experiments and had formulated a mathematical treatment. As early as September 18 and 25, 1820, he gave reports to the academy, confirming Ørsted's results and presenting some significant additional experimental results he had obtained himself. He also explained a hypothesis that he had formulated on the relationship between electricity and magnetism. His work during the next few years gave strong support to the conclusions he had reached so quickly. The book he published on the subject in 1827, *Memoir on the Mathematical Theory of Electrodynamic Phenomena,* is regarded as a masterpiece of style and clarity and has been called the *Principia* of electrodynamics; Maxwell later described it as "perfect in form and unassailable in accuracy." It is appropriate that the unit of electric current is called the ampere; it is one of the base units in the Système Internationale d'Unités (SI; see "A Few Points about Mathematics and Measurement" at the beginning of this book).

Ampère made careful studies of the effects of electric currents on one another and found that if currents traveled in the same direction along two parallel wires, there was repulsion between them; if the currents traveled in opposite directions, there was attraction. He also worked out a detailed mathematical treatment of the interactions on the basis of the assumption that current-carrying elements of the wires interacted with one another following the inverse square law. By adding up (using the mathematical procedure of integration) the effects of all the elements, he arrived at expressions that were consistent with the experimental results.

In carrying out these investigations Ampère developed a theory of magnetism, according to which it arises from electricity moving in circular orbits around the axis of the magnet (see fig. 23, *b*); that is, the planes of the orbits are at right angles to the magnetic axis. He

**Figure 23.** *a*, André-Marie Ampère (1775–1836); *b*, Ampère suggested in 1820 that a magnet can be created by an electric current flowing through a coiled wire, which he called a solenoid.

coined the word *solenoid* to refer to a wire wound in a spiral that creates a magnet when it carries a current. He carried out experiments with solenoids in which wires were wound around glass tubes and confirmed that when a current passed, a magnetic effect was obtained. He then developed an elegant mathematical treatment of the interactions between electric currents and the circular currents around the magnets and was able to explain Ørsted's results in terms of central forces.

Ampère's work was at once recognized by most investigators as an achievement of great importance. He later modified his theory to relate to the molecules in the magnets, suggesting that perpetual electric currents moved in orbits around them (today, in fact, we think of electrons moving in orbits about the nuclei of atoms). At the time there was no understanding of the nature of electricity; the electron was not to be discovered until over half a century later. Ampère also formulated a theory of the magnetism of Earth in terms of currents of electricity moving in orbits.

One who was particularly uncomfortable with Ampère's treatment was Michael Faraday. Since he knew little mathematics and was ill versed in physical theories, he simply could not understand the treatment. He could see nothing wrong with the idea of a circular force arising from a current flowing in a wire. Hardly anyone brought up on the physics of the time could accept such an idea, and yet it was the origin of Faraday's important concept of the electromagnetic field that was to be the core of the later ideas of Maxwell and many later scientists, including Heinrich Rudolph Hertz and Albert Einstein.

Michael Faraday (1791–1867; see fig. 24) was born to a Yorkshire family, his father, James Faraday, being a journeyman blacksmith. Since James was having difficulty making a living in the north of England, he moved the family to London early in 1771. Michael was born on September 22, 1771, so it is quite possible that Mrs. Faraday was pregnant during the uncomfortable trip to London. Michael was born at Newington Butts, which today is part of the London borough of Southwark, on the south of the River Thames, not far from London Bridge; the word *Butts* refers to the fact that the

**Figure 24.** Michael Faraday (1791–1867), one of the greatest experimental scientists of all time, made an important contribution by suggesting the idea of electric and magnetic fields, an idea developed by Maxwell and still of great value today.

land was at one time reserved for the practice of archery.

James Faraday could do little more than provide the basic necessities of life for his wife and four children. For some years the family of six lived in a single squalid room. Michael Faraday's parents imbued in him strong religious principles that were to dominate his life. The family belonged to a strict and somewhat obscure Christian sect, the members of which were known as the Sandemanians; in Faraday's own words, the sect was "despised." Although they were religious fundamentalists, the Sandemanians did not believe in a hell of fire and brimstone but placed more emphasis on love and a sense of social duty instead. Faraday remained a loyal Sandemanian throughout his life. Some idea of the strictness of the sect is provided by the fact that he was on one occasion disciplined for being absent from a religious gathering. His excuse, over which he himself had agonized for some time, was that he had been summoned to see Queen Victoria—but it was deemed inadequate.

Michael Faraday's education was rudimentary, consisting of little more than reading, writing, and simple arithmetic. At the age of thirteen he left school, and after delivering newspapers for a time he entered into a seven-year apprenticeship with a good-natured bookbinder, George Riebau, an émigré from France who had fled the excesses of the French Revolution. Riebau encouraged Faraday to read the books he was binding, and an article on "Electricity" in the *Encyclopedia Britannica* seems to have first aroused his interest in science. Also, the book *Conversations on Chemistry* made a particular impression on Faraday and perhaps did more than anything else to make him enthusiastic about science; its author was Jane Marcet (1769–1858), herself a chemist of some distinction, and throughout his life Faraday spoke enthusiastically about her book. In 1812 one of Riebau's customers gave Faraday a ticket to attend some of Sir Humphry Davy's lectures at the Royal Institution. In October 1812, when his apprenticeship expired, Faraday applied to Davy for a position, at the same time sending him notes he had prepared on his lectures. Davy was impressed but at first could do nothing. In March, 1813, however, when an assistant was dismissed for brawling, Faraday was appointed to his position at the Royal Institution and was given two rooms on the top floor in which he and his wife could live. The building was to be both his home and laboratory for many years.

Later that year Sir Humphry and Lady Davy paid an extended visit to France and Italy, and Faraday was invited to accompany them as scientific assistant. Although the experience was somewhat marred for Faraday by the fact that Lady Davy tended to treat him as a valet, he greatly profited from it, particularly from the opportunity of meeting many of the leading scientists of the time, including Volta, Arago, and Count Rumford.

At first Faraday's main work at the Royal Institution was in chemistry. By 1820 he had established something of a reputation as an analytical chemist, and from this work he was able to supplement his modest salary and even to contribute to the support of the Royal Institution, which was not well endowed. Some of Faraday's chemical work was concerned with clays, and some was on metal alloys; he prepared several novel varieties of steel, from some of which he fabricated razors for himself and his friends. Occasionally Faraday appeared in court as an expert witness. One of these cases related to oil, and

Faraday was led to investigate the properties of oils and gases, which were beginning to be used for public heating and lighting. It was as a result of these studies that Faraday discovered benzene in 1825, a substance of great importance. He also discovered numerous other important chemical compounds. Faraday was also the first to liquefy a number of gases, such as ammonia and carbon dioxide.

In 1820 Faraday began to look into the subject of electromagnetism. He repeated Ørsted's experiments and noticed that when a small magnetic needle was moved around a wire carrying a current, each of the poles turned in a circle. This caused him to speculate that a single magnetic pole, if it could exist, would move continuously around a wire as long as the current flowed. This led him to perform an experiment of great simplicity and also of great importance. In 1821 he attached a magnet upright to the bottom of a deep basin and then filled the basin with mercury so that only the pole of the magnet was above the surface. A wire, free to move, was attached above the bowl and was dipped into the mercury (see the right-hand bowl in fig. 25). When Faraday passed a current through the wire and the magnet, the wire continuously rotated around the magnet. In an adaptation of the experiment (see the left bowl in fig. 25), the magnet was rotated around the wire. He also succeeded in rotating

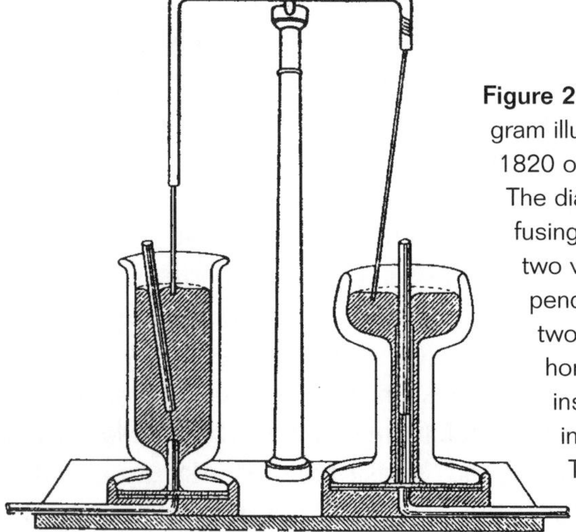

**Figure 25.** Faraday's published diagram illustrating his demonstration in 1820 of electromagnetic rotation. The diagram as drawn is a little confusing until one realizes that the two vessels are electrically independent and that they relate to two separate experiments. The horizontal bracket at the top is of insulating material, simply holding the electrodes in position. This was the first time that electricity had been converted into useful energy.

a wire by use of the magnetism of the earth. The great significance of these simple demonstrations is that electrical energy was being converted into mechanical energy for the first time. To Faraday the results implied that there were circular *lines of force* around the current-carrying wire, and he accepted this as a simple experimental fact. Almost everyone else concluded that the force could not be of this type but must be explained in some way in terms of central forces obeying the inverse square law.

At about the same time that Faraday was doing this work, electromagnets began to be constructed. It was found that iron filings left near a wire carrying an electric current became magnetized and that a steel needle became permanently magnetized if a current was passed through a coil of wire wrapped around it. A particularly significant observation was made in 1823 by William Sturgeon (1783–1850), who found that whereas a bar of steel could be magnetized permanently, a bar of soft iron was magnetized only temporarily; with soft iron the magnetism was lost as soon as the current no longer flowed. Sturgeon suggested the word *electromagnet* for a device in which magnetism is brought about by an electric current. Great improvements to electromagnets were made by distinguished American physicist Joseph Henry (1797–1878; see fig. 26). Henry was born in Albany, New York, to a poor family but unlike Faraday was able to get a good education. In about 1826 he was appointed professor of mathematics and natural philosophy at the Albany Academy. He remained there until 1832, when he was appointed professor at the College of New Jersey (now Princeton University). In 1846 he was appointed the first secretary of the Smithsonian Institution, which had just been established in Washington as the result of a bequest by eccentric Englishman James Louis Macie Smithson (1765–1829). Smithson was the illegitimate son of the first duke of Northumberland, and in a fit of pique at the rejection of a paper he submitted to the Royal Society, he left all his money to be used for the foundation of an institution in a country he had never visited.

Joseph Henry's interests in electricity and magnetism were similar to those of Faraday, but he was not quite as capable or persistent as Faraday, and time and again his discoveries came a little too late

**Figure 26.** American scientist Joseph Henry (1797–1878), made important discoveries, particularly in electricity and magnetism. Some of his contributions led to the development of electromagnets and the electric telegraph. (Edgar Fahs Smith Collection, University of Pennsylvania Library)

for him to be entitled to priority. His improvement of Sturgeon's technique for creating an electromagnet, carried out in 1826 when he was in Albany, was to insulate the wires before winding them onto the soft iron core. This allowed many more turns and therefore an increase in the magnetism produced. It is said that, unable to afford other material, Henry used his wife's petticoat, and perhaps her wedding dress, for the insulation of his wires. Henry also made the important discovery that the polarity of an electromagnet could be reversed by reversing the direction of the current. He continued to improve his electromagnets, and by 1832 he had constructed one that could lift a mass of over three thousand pounds. It is still in existence at the Smithsonian Institution. Electromagnets led, by only a short step, to the electric telegraph and the electric motor.

In 1831 Michael Faraday made one of his most famous discoveries, *electromagnetic induction*. He had often experimented with the idea in his mind that a wire bearing a current might induce a current in a nearby wire, but until August 29, 1831, he always failed to find any effect. On that day he wound one side of a soft-iron ring with insulated wire and arranged a secondary winding, connected to a galvanometer, around the other side (see fig. 27, *a*). At first he thought that the experiment was again a failure. He noticed, however, that when he *turned off* the electric current in the primary coil, believing the experiment to have failed, the galvanometer revealed a sudden short flow in the secondary circuit. Closer investigation showed that a *continuous* current in the primary circuit had no effect; it was only when the current was started or stopped that there was an effect on the galvanometer.

Soon after his discovery of electromagnetic induction in 1831, Faraday demonstrated that if he pushed a magnet into a coil of wire, a transient current was produced (see fig. 27, *b*). When the magnet was withdrawn a transient current was generated in the opposite direction. No current passed when the magnet was stationary; to generate a current the magnet must be in motion relative to the coil.

It is important to appreciate that electromagnetic induction only occurs if there is a *change* in an electric current, which means that there is a *change* in magnetism. Ørsted and Ampère had shown that a *steady* electric current produces a magnetic field, and investigators thought at first that if a wire were placed near a magnet, an electric current would be generated in it. In hindsight we can see today that such a result would be impossible. We are now aware of what Faraday was not at the time—of the necessity for energy to be con-

**Figure 27.** Two diagrams from Faraday's diary of 1831: *a*, illustration of his discovery of electromagnetic induction—when he caused a current to flow in coil A, called the primary coil, Faraday observed a short-lasting current in coil B, the secondary coil; *b*, illustration of an experiment in which a magnet induces an electric current in a coil when a magnet is pushed into, or pulled out of, the coil.

served. If a stationary magnet (either a permanent magnet or an electromagnet) were to induce an electric current in a nearby wire we would be getting something for nothing: where would the energy have come from? Ørsted had discovered that a *steady* current of electricity would affect a magnet, but we now know that an electric current involves a *moving* electric charge and that energy (from a battery, for example) is being used up in creating it. Investigators had also found that an electric current (a *moving* electric charge) in a coiled wire would produce a magnet. But we cannot produce an electric current from a stationary magnet; we need some *change* in a magnetic field, which can be produced by *moving* a permanent magnet (which uses energy) or by switching on or off the current in a wire.

In 1832 Faraday began to devote much attention to electrolysis—the processes that occur when an electric current is passed into a solution and brings about chemical change. One of the results he was able to show from his experiments is that electrolysis can be brought about by electricity produced in a variety of different ways, such as from electrostatic generators, voltaic cells, and even electric fish. He showed that electrolysis can occur not only if a current passes but also if an electric discharge is passed through a solution, without any wires being introduced into it. Electrolysis does not, however, occur when a solution is subjected to an electric field, however intense; it is essential for a current to pass or an electric discharge to occur. Faraday performed one experiment in which two solutions were separated from one another by a seventy-foot string soaked in brine. Gases were evolved at the wires immersed in the solution, and he concluded that, for electrolysis to occur, it is necessary for a discharge to take place or for a current to flow. He also showed that the effects of electrolysis do not follow straight lines, which they would do if the inverse square law was obeyed. These results convinced him that electrolysis cannot be explained simply in terms of some kind of central force obeying the inverse square law. He concluded instead that electricity is carried through the solution from molecule to molecule.

In 1834 Faraday suggested two fundamental laws of electrolysis. It is often stated that he discovered these two laws empirically—that is, entirely on the basis of his experiments—but this is not the case; they were deduced by him on the basis of his ideas about how electricity interacts with a solution and were later confirmed by him through experimentation. According to Faraday's ideas, the amount of material deposited in an electrolysis would depend mathematically on the product of the current and the time, which is the *quantity* of electricity that passes through the solution, and Faraday confirmed that this is the case; this was his *first law of electrolysis.* His *second law of electrolysis,* which we need not discuss in depth, is concerned with the relative amounts of different substances that are evolved or deposited. Faraday's laws of electrolysis were of great theoretical significance, since they suggested that electricity itself is not continuous and that fundamental particles of electricity are in some way associated with atoms. Faraday, however, placed little emphasis on this interpretation, leaving it for others to develop.

▲▲▲

Much of Faraday's great work on electricity and magnetism had been completed by the time James Clerk Maxwell became a student at Cambridge in 1850. Most scientists at the time, though admiring of Faraday's experimental achievements, were scornful of his limited grasp of accepted physical theory and considered his ideas of circular fields of force to be absurd. The youthful Maxwell, however, thought otherwise and was one of the few to realize that Faraday's unconventional ideas were perfectly reasonable. Maxwell's first publication on the subject, titled "On Faraday's Lines of Force," appeared in 1856 when he was twenty-five and still at Cambridge. It was convincing in showing mathematically that Faraday's ideas of circular electric fields gave a valid alternative to Ampère's treatment based on central forces obeying the inverse square law. Maxwell was a modest man who always insisted that he did nothing more than express Faraday's ideas in mathematical form; in reality he did much more than that. Producing the mathematical equations was far from a routine task that any highly skilled mathematician could have carried out; it also depended on clarifying and modifying the basic concepts. In spite of his great ability Maxwell had to struggle for many more years before arriving at a complete theory.

In the early 1860s, when Maxwell was at King's College, London, he published a series of papers that were mainly concerned with devising a model for the *ether*, the invisible medium that was assumed to pervade the whole of space and transmit light. His model was such that a changing magnetic field gave rise to an electric field and that a changing electric field produced a magnetic field. In the end Maxwell developed a theory of light, electricity, and magnetism that applied whether the ether existed or not. His complicated theory of the ether, although interesting, is not today regarded as of much importance, since there is no evidence that the medium exists. More significant to the outcome were theoretical and experimental investigations carried out by him and others at about the same time on some fundamental properties of electricity and magnetism.

These began as a result of a scientific committee set up in 1861 to establish a set of electrical and magnetic standards; committees usually produce little, but this one was an exception. It had been realized that critical insight into electricity and magnetism could be obtained by comparing results expressed in the two sets of units, *electrostatic* and *electromagnetic*, that are used in this field. The electrostatic units, particularly appropriate to static electricity, relate the force of attraction and repulsion between two charged bodies to the quantities of electricity that they hold. The electromagnetic units, on the other hand, are concerned with forces due to electric currents. The quantity of electricity residing on a wire at a given instant depends on the speed with which the current passes along the wire. It turns out that the ratio of the value of an electrical quantity expressed in electromagnetic units to that expressed in electrostatic units is the speed with which the electricity travels.

Maxwell was an active member of the committee and carried out experimental investigations in addition to doing theoretical work. His conclusion from the experiments comparing electrical quantities expressed in the two sets of units was that the speed of an electric current was close to 300,000 km/s (671.1 million miles per hour), which is the speed of light. He used another procedure, involving the electrical and magnetic properties of a vacuum, and came to the same conclusion: the speed with which electrical and magnetic waves travel is exactly the same as the speed of light.

Maxwell was justifiably impressed by this significant result. He was struck by the fact that he had been able to calculate the speed of light from the results of experiments on electricity and magnetism without light itself being in any way involved. In Maxwell's own words, "The only use of light in the experiments was to see the instruments." This work led to the inevitable conclusion that light, electricity, and magnetism have much in common. The same electromagnetic theory therefore applies to light, to an electric field, and to a magnetic field. The field produced by an electric current has not only an electric component but also a magnetic one, and the field produced by a magnet has an electric as well as a magnetic component. Light also has both electric and magnetic components.

A useful result of this work was that it led to a decision between the two alternative theories of electricity that were held at the time. One theory was that there were two electrical fluids, positive and negative, which moved in opposite directions when an electric current flowed. The method of comparison of units led to the conclusion that if there were two fluids, they would each flow with half the speed of light. The evidence thus supported the theory (which Benjamin Franklin, among others, had advocated) that there is only one type of electricity that flows. Today we know that a flow of electrons is involved.

Maxwell's final major paper on electromagnetic theory appeared in 1864. With uncharacteristic immodesty, but with perfect justification, he described this paper to a friend as "great guns." In this paper he ignored the rather elaborate and artificial model he had proposed for the ether and concentrated on the propagation of electromagnetic waves through space. The position he took, which is accepted today, is that the mathematical treatment remains valid without any assumptions about the nature of the medium through which the waves travel. In other words, he threw out the bathwater but carefully kept hold of the baby!

In doing so, Maxwell made an important break with scientific tradition. Previously it had been felt necessary to base a scientific theory on a model that could be clearly visualized. Maxwell's theory, on the other hand, represented the situation not in terms of a model but as a mathematical analog. Maxwell's friend and colleague William Thomson (better known today as Lord Kelvin, whom we shall meet later) always insisted on a mechanical model; in his own words, "I never satisfy myself unless I can make a mechanical model of a thing. If I can make a mechanical model I can understand it." As a result, Kelvin never really appreciated Maxwell's theory.

The theory can be expressed in terms of a few equations that have been referred to as "simple." They are indeed simple to look at, but understanding them demands a considerable background knowledge of electrical and magnetic theory and of vectors. Maxwell's famous book *Treatise on Electricity and Magnetism* was published in 1873, when Maxwell was back at Cambridge as its first Cavendish Professor.[6] The book expounded and further developed his theory of electromagnetic radiation; a significant feature of it is that the word *ether* is mentioned only once and that the model for the ether elaborated in his 1861–62 paper is referred to only incidentally.

Maxwell became Cavendish Professor at Cambridge in 1871, and the circumstances of his appointment are interesting. The post was first offered to his friend Thomson, who had established a great reputation as professor of physics at the University of Glasgow; he was unwilling to leave Glasgow, where he remained until his death in 1907. The post was then offered (at Thomson's suggestion) to distinguished German physiologist and physicist Hermann von Helmholtz; but he also declined, since he had just accepted the chair at Berlin.

As third choice Maxwell, who had just reached the age of forty, was offered the position, which he accepted reluctantly. He was probably the greatest scientist of the three but perhaps not the best choice as professor; Thomson and Helmholtz were both inspiring men and excellent lecturers and would have attracted large classes. Maxwell's classes were pitifully small, averaging between two and three students. He never enjoyed lecturing and was not good at it. His Scottish brogue was strong, his diction was poor, and he had something of a stutter when he was nervous; as a result, he could not hold the attention of his students. He also had an unfortunate tendency, throughout his career, to make mathematical mistakes. He was even apt when lecturing to express his ideas in a confused way. For a man who wrote with such great clarity, this is most surprising; even though he was conscientious in preparing his lectures he became muddled when he started to speak. He made up for these deficiencies by attracting a number of excellent research students; his contributions to the design of the Cavendish Laboratory were very effective, and he personally arranged for the construction of much of the apparatus to be used in the practical classes, paying for some of it out of his own pocket.

Maxwell accomplished a great deal during his eight years at the Cavendish. Unfortunately his term at the laboratory was short, as he became fatally ill in the spring of 1879; in October he was told that he had abdominal cancer (from which his mother had died at the same age) and would not live more than a month or so. Despite great pain he continued research almost to the day of his death on November 5, at the age of forty-eight. There was little publicity at the time of his death, and it seems likely that his friends let it be known that because of his modesty and unpretentiousness, any public recognition would be inappropriate. He was buried in the churchyard of the tiny Parton Kirk near his manor house at Glenlair. For many years his grave was hard to find, but a brass plate installed in 1989 now describes him with apt simplicity as "A Good Man, Full of Humour and Wisdom."

▲▲▲

Maxwell's electromagnetic theory had many theoretical and practical consequences, which we can touch on only briefly.[7] Perhaps the most important of these resulted from the investigations of Heinrich Rudolph Hertz (1857–94; see fig. 28), a remarkable man who accomplished much in his brief life of thirty-six years. He was born in Hamburg into a prosperous and cultured family and at first prepared for an engineering career. Later he began to favor pure science and studied at the universities of Munich and Berlin. Later, at the University of Karlsruhe, he carried out experiments on the transmis-

**Figure 28.** German physicist Heinrich Rudolph Hertz (1857–94) in 1887 first confirmed Maxwell's theory of electromagnetic radiation by transmitting radiation over a short distance and confirming that it traveled with the speed of light.

sion of radiation that were to make him world famous. He produced electromagnetic radiation by producing an electric spark, 5 cm (about 2 in) in length, between two brass knobs and found that when a spark passed across a spark gap, a spark was produced in another spark gap a few yards away. It was this experiment that made possible the extensive radio and television transmission that plays such a great role in modern society.

These applications are popularly attributed to Guglielmo Marconi (1874–1937), who made important contributions to making radio commercially possible but did little if anything that was scientifically or technically new. The credit for first transmitting a radio signal should instead go to British physicist Oliver Lodge (1851–1940). His first demonstration of radio was before a large audience at the Royal Institution in London on March 8, 1889. His most famous demonstration was made on August 14, 1894, at a meeting in Oxford of the British Association for the Advancement of Science. The demonstrations made by Marconi several years later were with equipment similar to that used by Lodge.

In honor of Hertz his name is used for a unit of frequency; 1 hertz (1 Hz) is the same as 1 reciprocal second (1 $s^{-1}$). The unit Hz must only be used for frequency in the sense of cycles per second; the unit

$s^{-1}$ used in any other sense should not be written as Hz (e.g., the unit *meter per second,* m · $s^{-1}$, for speed should not be written as m · Hz).

The work of Hertz greatly extended our knowledge of the electromagnetic spectrum. Before he did his work, not much more than the visible spectrum was known. In 1800, famous astronomer William Herschel discovered, on account of its heating effect, that there was some invisible radiation beyond the red, called the *infrared.* The following year German scientist Johann Wilhelm Ritter discovered *ultraviolet* radiation, from the fact that it blackened silver chloride. There was, before Hertz did his experiments, knowledge of only a little more than the visible spectrum. The wavelength range was hardly more than a factor of two.

Figure 29 shows a version of the electromagnetic spectrum as it is

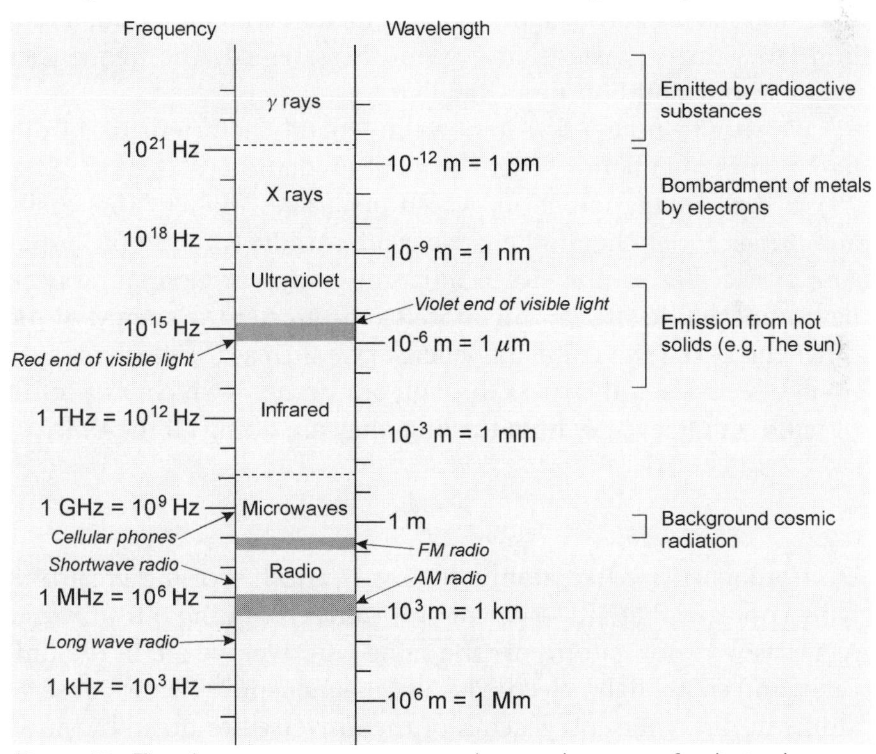


**Figure 29.** The electromagnetic spectrum, showing the types of radiation known today and some of the ways in which different kinds of radiation are produced. (The prefixes commonly used with wavelengths and frequencies can be found in "A Few Points about Mathematics and Measurement," p. 17.)

known today. It includes x-rays, which were discovered in 1895 by German physicist Wilhelm Konrad Röntgen (1845–1923), and gamma ($\gamma$) rays, emitted by radioactive substances, which were discovered in 1900 by French physicist Paul Ulrich Villard (1860–1934).

The figure shows the wavelengths and frequencies associated with the various types of radiation. We see that visible light covers only a tiny fraction of the spectrum known today. Red light has a wavelength of about 700 nanometers ($7 \times 10^{-7}$ m, or 7,000 Å). The frequency corresponding to it is about 400 million million Hertz, or 400 terahertz (400 THz). Violet radiation has a wavelength of about 400 nanometers (4,000 Å), or roughly half that of red radiation; the frequency is therefore roughly twice that of red radiation. What is remarkable about Hertz's contribution was that he had discovered radiation having wavelengths of a few meters, which is about a million ($10^6$) times greater than previously observed. The frequencies are therefore a million times smaller.

We should note a few important general characteristics of different types of radiation. High-frequency radiation (at the top of fig. 29) is high-energy radiation, which penetrates matter more easily and brings about chemical change more readily. For example, exposure of the skin to ultraviolet radiation, x-rays, or gamma rays can have undesirable effects—the more so the greater the energy. At the other end of the spectrum, low-energy rays such as radio waves cause no ill effects when they pass through our bodies—which, as a result of radio and television broadcasting, they are doing all the time.

▲▲▲

Electromagnetism, like atomic theory, is another of the great unifying concepts that are so essential for understanding our universe. As far as we know, atoms are the same wherever we are in the universe, and so are light, electricity, and magnetism. When we observe atoms in the outermost reaches of the universe, we do so by virtue of the electromagnetic radiation they emitted many billions of years ago. Every atom in the universe has electromagnetic radiation associated with it and exhibits electrical and magnetic properties.

Another of the unifying concepts relates to the energy of the universe. Our discussions of atoms and electromagnetism may have left the impression that the universe is static, but we know that this is far from the case. The universe has changed drastically from its creation in the big bang to what it is today. The key to what has happened is provided by a study of energy, which has many aspects. There are several forms of energy, and thanks to Einstein's great insight we know that matter and energy are interchangeable. Everything that happens, and how fast it happens, is controlled by energy. In the next chapter we shall try to gain some insight into the unifying role of energy.

3

# THE WORLD OF ENERGY

Today we have become so accustomed to using energy and paying for it that we all have a fairly good idea of what it is. We know that heat can be converted into work, and vice versa, and that both are forms of energy. Before the middle of the nineteenth century, however, these ideas were still a matter of confusion, and the concept of energy was not even recognized. Even the word "energy" is a comparatively new one; Isaac Newton, who did so much pioneering work in mechanics, never seems to have used it.

It is not easy to give a precise definition of energy in simple language. We can say that energy is either work or anything that can be converted into work. By work we mean something that is done whenever there is movement of a body against a resisting force. If there is a heavy piece of furniture on the floor and we push against it but fail to move it, we are doing no work on it, even though we may be perspiring. If we push harder and successfully move it, we are then doing work against the resisting force of friction. If in moving something we exert a constant force and push it through a certain distance, the scientific definition of the work is that it is equal to the force multiplied by the distance.

The reason that the concept of energy was rather elusive is that, unlike other physical quantities such as pressure, volume, and temperature, we cannot observe it directly. We must always infer the energy of an object in some way, for example from its temperature or its capacity to do work.

The question of the different forms of energy is also a little confusing, particularly as there are several systems of classification, and some of the terms have been used in different ways. From the standpoint of Newton's mechanics there are just two forms of energy: kinetic and potential. Kinetic energy is energy that a body has by virtue of its motion, and for a body of mass $m$ and speed $v$ it is defined as $\frac{1}{2}mv^2$. A body that is not moving has no kinetic energy, and we note that the kinetic energy depends on the square of the speed. A body that moves twice as fast as another of the same mass thus has four times the kinetic energy.

Potential energy is energy that a body has by virtue not of how it moves but of where it is. A special type of potential energy is *gravitational energy,* which results from the force of gravity. If a body of mass $m$ is raised from the ground to a height $d$, we say that its potential energy or gravitational energy has increased by $mgd$, where $g$ is a factor known as the *acceleration of gravity*. The product $mg$ is actually the weight, or force, exerted by the mass when it is subjected to the gravitational attraction of Earth.[1] This potential energy $mgd$ is equal to the *work* we have to do in raising the body to that height $d$. An electrically charged body also has potential energy when it is in an electric field.

Potential energy arises as a result of a force. A body is attracted to Earth by the force of gravity, and it requires work to raise a stone to a greater height. This is because Earth attracts the stone more at the lower height than it does after it has been raised. A stone that has been taken to a region of space where there is negligible gravity can be moved without any work being required; the potential energy is the same before and after the move. This shows us that a force is exerted only when there is a variable potential energy. The gravitational force, for example, is due to the fact that the gravitational energy varies with the height from the ground.

There are other ways of classifying energy. For example, it is often convenient to speak of electrical energy, chemical energy, and nuclear energy, but like gravitational energy, these for the most part are special types of potential energy. Sometimes we speak of the internal energy of a system, and this is a little more complicated since it is a combination of potential energy and the kinetic energy of the atoms contained in the system.

It was realized even before energy had been precisely identified that it could not be created from nothing and could not be destroyed. Many devices, some of them highly ingenious, have been invented that were supposed to go on operating forever, without help from any outside agency. Such a hypothetical device is now referred to as a *perpetual motion machine of the first kind,* and it became clear that such machines never work. As early as 1775 the French Académie des Sciences passed a resolution that they would no longer consider any machine claiming to exhibit perpetual motion.

For many years scientific investigators had different ideas about the nature of heat. In the early nineteenth century the most popular theory was that proposed by distinguished French chemist Antoine Lavoisier (1743–94). He had carried out investigations that seemed to show that heat was a substance of a rather special kind in that it had no weight. Lavoisier referred to it as an "imponderable fluid" and even included *calorique* in his list of the chemical elements. Since Lavoisier was held in great esteem—and since most of his other ideas were certainly correct—his idea that heat is a substance was taken seriously by many eminent scientists of his time. The experiments of Scottish chemist Joseph Black (1728–99) at the University of Glasgow and later at the University of Edinburgh also seemed to lead to the conclusion that heat is a substance.

The alternative—and as we now know, correct—idea was that heat results from motion. This was the view of many distinguished people, even much earlier. Francis Bacon (1561–1626), who though primarily a lawyer and statesman had thought much about science and technology, produced strong arguments for believing that heat is not a substance but a form of energy. A similar view was taken by Robert Boyle (1627–91) as well as Newton. Both of them regarded matter as composed of atoms, and their theory in modern terms was that heat was the kinetic energy of the atoms. This idea was expressed very clearly by philosopher, physician, and investigator

John Locke (1632–1704): "Heat is a very brisk agitation of the insensible parts of the object [i.e., of the atoms], which produces in us that sensation from which we denominate the object hot; so that what in our sensation is heat, in the object is nothing but motion."

Reaching the correct conclusions about the nature of heat and its relationship to mechanical work was by no means a simple matter. Some help came from purely technical and empirical work on the development of steam engines. The steam engine was invented, and indeed brought to a high degree of perfection, by men who had no training in science and sometimes little knowledge of it. Only after a particularly efficient steam engine had been built, in the latter part of the eighteenth century, did scientists begin to investigate how it worked. Their conclusions were embodied in the two basic laws of thermodynamics. These laws, so important in all of science today, thus owe a good deal to people who were not seeking the truth for its own sake but were extremely ingenious engineers.

The most distinguished of these was James Watt (1736–1819; see fig. 30). He was born in Greenock, Scotland, and after learning the trade of instrument maker in London he practiced his trade for many years at the University of Glasgow, where he soon established a reputation for ingenuity and persistence. At the time the professor of natural philosophy (i.e., physics) at Glasgow was John Anderson (1726–96), who had a model of a steam engine that had been devised by Thomas Newcomen (1663–1729). It worked very badly, and Anderson asked Watt to overhaul it. Watt went about this task with great persistence and carried out numerous investigations on the thermal effects of mixing steam and water. While working on the model of the Newcomen engine, Watt realized that it had a fundamental flaw. The cylinder in it was first heated with steam and then cooled with water, obviously involving much unnecessary wastage of heat and loss of efficiency. Watt then had the most important of his many innovative ideas, the *separate condenser*. His idea was that there should be two cylinders connected together: one always kept hot and the other, the condensing cylinder, always cold. Watt's engines were much more efficient than earlier ones, and they found many applications. The first locomotive engine was built in 1784 by

**Figure 30.** James Watt (1736–1819), distinguished British engineer and scientist, famous for his highly original work on steam engines. (Edgar Fahs Smith Collection, University of Pennsylvania Library)

William Murdoch (1754–1839), Watt's assistant. He built a model locomotive in Cornwall, and since there were no railway lines on which to test it, he made use of the steep banks of a lane leading to the local church. When he made his first trial, at night, the engine moved so fast that it outdistanced him and almost ran down the rector of the church, who happened to be walking along the road. Running away in terror, the rector assumed the fiery monster to be the embodiment of the devil.

Watt had an excellent understanding of the operation of steam engines, but he hardly concerned himself with just what heat is. On the whole he tended to accept the view of his friend Joseph Black that heat is a substance. Nothing in the operation of a steam engine seemed to point strongly toward one particular explanation of heat. Heat can be a form of motion, which in an engine is converted into the energy of the pistons; this we now know to be correct. Alternatively, the operation of an engine may be thought of by analogy with a waterfall causing a wheel to turn (see fig. 31). It can be supposed that the heat is a substance that flows from a higher to a lower temperature, just as water falls from a higher to a lower level. When water causes a water wheel to turn, there is no loss of water, the wheel being turned by the force of the water. In the same way, the force of falling heat could cause the piston in an engine to move and perform work, without any loss of heat. It was not for several decades that this second explanation was completely discarded.

▲▲▲

One of the most challenging scientific problems that was solved in the nineteenth century was the precise relationship between heat and mechanical work, a field of investigation that became known as *thermodynamics*. The work on the functioning of steam engines, particularly the investigations made by James Watt, played an important

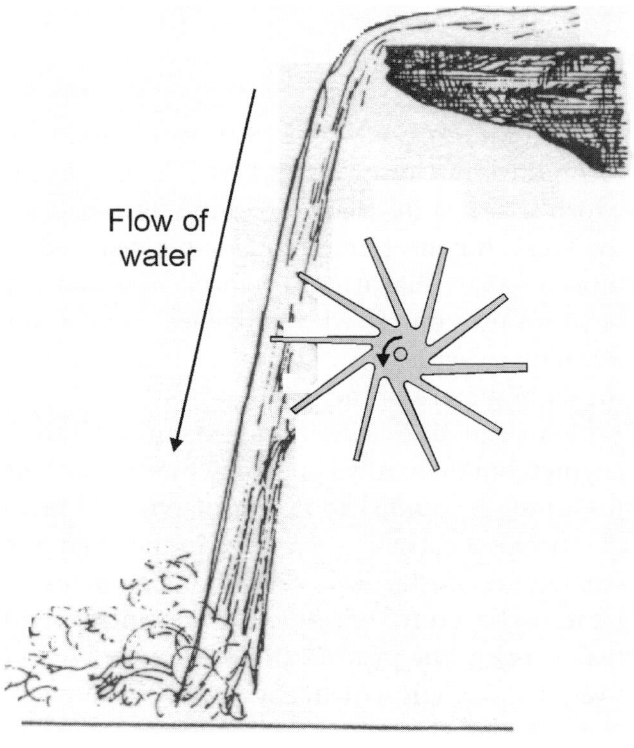

Flow of water

**Figure 31.** Falling water can perform work by turning a water wheel. It was at one time thought that a steam engine operates in a similar way, the work being produced simply as a result of heat falling from a higher to a lower temperature. We now know that some of the heat is actually converted into work.

part in this study, but much else remained to be done, and several people made contributions. It turned out that there are two basic principles, summarized by the first two laws of thermodynamics. The first law is closely related to the principle that energy is conserved; it cannot be created or destroyed. Three men played major roles in establishing this law. One of them was American-born Benjamin Thompson, who later became Count Rumford. He was impressed by the heat produced in the boring of cannons and concluded that it must be produced by the work done in the process. Another was German physician Julius Robert Mayer; he did hardly any experiments but pondered digestive processes as well as other work done by humans. The most detailed and convincing experiments that proved energy to be conserved were done by James Prescott Joule, an English amateur scientist. Beginning in about 1837 he carried out a variety of careful experiments, and his investigations completely transformed the subject. We will look at their work in more detail.

Benjamin Thompson (1753–1814; see fig. 32) was born in Woburn, Massachusetts, of parents who kept a small farm. He received an adequate education and at the age of fourteen was apprenticed to a storekeeper, but the work was not to his liking, and he soon abandoned it. Later he boarded with wealthy families, acting as a tutor to their children. At the age of nineteen he caught the eye of a wealthy widow, Mrs. Pierce, who was fourteen years his senior, and within four months of their first meeting they married in November 1771. They settled at his wife's home in Concord, New Hampshire, which had formerly been called Rumford. This marriage at once converted him into a wealthy, landed gentleman, and he took every advantage of his improved circumstances. Throughout his life he had a sharp eye for the main chance and within a short time was granted a major's commission in the Second Provincial Regiment. Relations with Britain were at the time strained, and Thompson favored the British side. In 1776, on the outbreak of the War of Independence, he fled to England, abandoning his wife and baby daughter. He never saw his wife again and had little correspondence with her. Much later he resumed contact with his daughter.

In England he gave valuable information to the government on

**Figure 32.** Benjamin Thompson (1753–1814), later Count Rumford, was born in Massachusetts, but most of his scientific work was done in England and Bavaria. He was the first to present convincing evidence for the conservation of energy and made many practical innovations. (Edgar Fahs Smith Collection, University of Pennsylvania Library)

the situation in America (in other words, he acted as a spy) and became acquainted with many prominent people, including King George III, who at first was friendly with him but, like most people, soon became disillusioned. Thompson had his portrait painted in full military regalia by the great Thomas Gainsborough. He carried out some experiments with gunpowder and as a result was elected a Fellow of the Royal Society in 1781, at the young age of twenty-seven. He was later commissioned a colonel of the Regiment of Artillery and soon retired with a lifetime pension.

In 1784 he was knighted by King George and left England to enter the service of the Elector of Bavaria, with whom he had become acquainted in his characteristic style. At the time, Bavaria was in a sad condition, and in a short time Col. Sir Benjamin Thompson had brought about a number of remarkable improvements. He reformed the army, established a military academy, planned a poor-law system, spread the knowledge of nutrition and domestic economy, and improved the breeds of horses and cattle. He arranged for the construction of a large workhouse into which beggars were rounded up and required to work at making military uniforms, at the same time being housed and fed and generally well treated. For these services Thompson was made a major general in the Bavarian army, Minister of War, Royal Chamberlain, a privy counselor, and chief of police (Gilbert and Sullivan's Pooh-Bah did not do much better). He was also created a Count of the Holy Roman Empire, choosing the title Count Rumford, the former name of the town of Concord, where he had lived.

He decided to upgrade the artillery and in particular to arrange for the manufacture of heavy brass cannons. He personally supervised the boring of the cannon barrels and was impressed by the fact that much frictional heat was continuously produced. He decided to make measurements of the amount of heat produced and for this purpose arranged for the casting of a specially shaped cannon barrel that could be insulated against loss of heat. He replaced the sharp boring tool with a dull drill to increase the friction and thus generate more heat. By immersing the drill in a tank full of water and making temperature measurements he was able to determine the amount of heat produced.

He was impressed by the fact that, as he said in his report to the Royal Society, "at 2 hours and 30 minutes it ACTUALLY BOILED. It would be difficult to describe the surprise and astonishment expressed in the countenances of the bystanders, on seeing so large a quantity of cold water heated, and actually made to boil, without any fire." He continued, "Though there was, in fact, nothing that could justly be considered as surprising in this event, yet I acknowledge fairly that it afforded me a degree of childish pleasure, which, were I ambitious of the reputation of a *grave philosopher*, I ought most certainly rather to hide than to discover."

Rumford was particularly surprised by the lavishness with which the heat was produced. He established that no change of mass occurred during the process and that the metal shavings had the same properties as the metal before it was bored. From this he concluded that heat could not be a substance, a conclusion he expressed clearly and convincingly: "anything which any *insulated* body . . . can continue to furnish *without limitation*, cannot possibly be a *material substance*." Heat, he concluded, cannot be explained "except it be Motion." In his cannon-boring experiments there was thus a conversion of mechanical work into heat, and he obtained a value for the "mechanical equivalent of heat," which is the amount of work required to produce a given amount of heat.

Although the results of these experiments were impressive, they by no means convinced most scientists that heat is a form of motion rather than a substance. The skepticism with which Rumford's work was received may have been largely due to his personal unpopularity arising from his general behavior, which most people found insufferable.

In 1799 Rumford left the Bavarian service and returned to England. He was intimately concerned in 1799 with the founding in London of the Royal Institution of Great Britain, which continues to this day to do outstanding work in a variety of fields. However, his association with the Royal Institution was short-lived, as he soon quarreled with its managers. In 1804 he moved to Paris, where he renewed an acquaintance with Lavoisier's wealthy widow, whose husband had been guillotined eleven years earlier. They were married in October 1805; Rumford was fifty-two and his wife, from whom he had been separated for thirty years, had died; Madame Lavoisier was forty-five. But they were two strong-willed people and soon separated.

Rumford's scientific and organizational achievements were outstanding. One of his inventions, the so-called Rumford fireplace, brought him considerable wealth. It warmed a room much more effectively than a conventional grate and greatly reduced the escape of smoke into a room; modern fireplaces are based on the principles he established. In connection with the workhouse in Munich he designed very effective lamps and kitchen ranges that could be used for feeding large numbers of people. In his work with the Bavarian army he performed useful experiments on the effectiveness of various types of clothing in keeping people warm, and his findings were of considerable practical importance.

Rumford's personal relationships usually soured rapidly, since his treatment of people was complex and inconsistent. On the one hand, he loved humanity in the abstract and put tremendous effort into easing the lot of the disadvantaged. On the other hand, his behavior toward those with whom he came into personal contact was arrogant and overbearing. His own assessment of himself is rather astonishing: "No man supported a better moral character than I do, and no man is better satisfied with himself."

▲▲▲

Several decades later, in the middle of the nineteenth century, two men played a particularly persuasive role in convincing everyone that heat is a form of energy and that energy is conserved: German physician Julius Robert Mayer and English brewer and amateur scientist James Prescott Joule.

Julius Robert Mayer (1814–78; fig. 33) was born in the year of Rumford's death, in the south German town of Heilbronn. At university he studied medicine but little physics, which was a serious disadvantage to him in his later investigations. He passed his state medical

**Figure 33.** German physician Julius Robert Mayer (1814–78), who in 1842 produced convincing arguments, based on physiological evidence, in favor of the conservation of energy.

examinations in 1840, and in order to see something of the world he signed on as doctor on the Dutch merchant ship *Java,* which sailed out of Rotterdam bound for the Dutch East Indies. When the ship was off Indonesia, some of the crew succumbed to an epidemic, and following the practice of the time he performed a number of bloodlettings from veins in the arm. When he did so, he noticed that the blood was unusually red, as if it had come from an artery instead of a vein.

To explain his observation that venous blood is redder in the tropics than in a colder climate he suggested that in warmer weather there is a lower metabolic rate, less oxidation being required to keep the body warm. As a result of the smaller consumption of oxygen there is less change in the color of the blood and less contrast in color between venous and arterial blood. This caused Mayer to think deeply about the relationship between food consumption, heat production, and work done. He also carefully studied some of the heat studies made by others but did no experiments on the subject himself.

On the basis of evidence of this kind, Mayer arrived at the conclusion that heat and work must be interconvertible. On his return home in early 1841 Mayer prepared a scientific paper on his ideas. Unfortunately, it was couched in religious and metaphysical terms, and since he was ignorant of mechanics he made many elementary errors. He also used personal jargon that was inappropriate for a scientific paper. In June 1841 he submitted it to the leading German journal of physics, the *Annalen der Physik und Chemie,* but it was rejected. Mayer was angry and frustrated at this treatment, but it had the effect of stinging him into action. He soon agreed that the paper did have serious limitations and revised and improved it extensively, making it much more convincing. Nine months after submitting the paper for the first time he submitted it again but instead to the *Annalen der Chemie und Pharmacie.* It was at once accepted.

Mayer later claimed with some justification that this paper established his priority for the principle of conservation of energy. Even in its revised form, however, the paper was not an impressive one, as it relied a good deal on rather casual observations and intuitive impressions, and hardly at all on the results of carefully planned experiments. Because of this, and since he did not belong to the scientific elite of the time, his ideas were at first either ignored or ridiculed. Mayer was a proud and sensitive man, and this treatment had a serious effect on his mental stability. In 1850, overcome by the strain, he attempted suicide by jumping out of a third-floor window and falling nearly ten meters (thirty-three feet). He survived but suffered severe injuries, which caused a further deterioration of his mental condition. Several times he was confined in mental institutions and sometimes had to be restrained in a straitjacket. His life, however, had a happier ending, since there was final recognition of his accomplishments. He was elected a corresponding member of the French Académie des Sciences, received some honorary degrees, and in 1871 received the Copley Medal of the Royal Society of London.

The work done by James Prescott Joule (1818–89; see fig. 34) was of a very different character from that of Mayer, since it involved carefully designed and executed experiments that left no doubt about the interconversion of heat and work. A member of a prosperous brewing family, Joule was born in Salford, near Manchester, and received much of his education at home. Joule never held any academic or research appointment. All his investigations were carried out in laboratories established at his own expense and installed both in the brewery and later in his various homes.

**Figure 34.** British amateur scientist James Prescott Joule (1818–89) performed careful experiments from 1837 to 1847 that established beyond question the principle of conservation of energy.

From 1837, when he was nineteen, until about 1847, Joule carried out investigations that were particularly focused on the interrelationship between heat and work. His most important investigations were inspired by some recent work on electromagnets and electric motors. He was also influenced by an idea that had become prevalent, but which he helped to disprove, that there was no limit to the power that could be obtained from a motor operated by an electric battery. As mentioned earlier, batteries for the generation of electric current were developed in the first decades of the nineteenth century. Steam engines were usually supplied with coal obtained from under the ground, and it was realized even in the early nineteenth century that Earth's supply of coal was limited. Electric batteries, on the other hand, led to what has been called an electrical euphoria. It was argued that if certain imperfections of the electric motor could be eliminated, a motor would go on accelerating indefinitely, performing enormous amounts of work. These arguments, though now known to be fallacious, seemed at the time compelling.

At first Joule was also somewhat carried away by this electrical euphoria, with its promise of creating energy. Beginning in 1837 he carried out careful experiments on the mechanical effect that could be obtained from an electric motor and related it to the amounts of metal used up in the battery operating the motor. He was disappointed to find that the consumption of a given amount of zinc in a battery would lead to the production of only about one-fifth of the mechanical work that would be produced by the same weight of coal in a steam engine. Since zinc was much more expensive than coal, this meant that an electric motor was far from being competitive with a steam engine for the primary production of energy.

Joule then decided to study the heating effect of an electric current. He soon concluded that the heat produced is equivalent to the energy released by the chemical action occurring in the battery, and that energy could not be created from nothing. With regard to the applications of electricity to practical use, Joule wrote that "electricity is a grand agent for carrying, arranging, and converting chemical heat." That was a shrewd prophesy; this is exactly what modern batteries do.

In 1843 Joule enclosed the revolving part ("armature") of an electric generator (sometimes later called a "dynamo") in a vessel containing water and determined the heat generated when he rotated the armature for a fixed period of time. He also measured the heat produced by the electric current that was generated. In this way he established the equivalence between the heat produced as a result of rotating the armature and the mechanical work required for the rotation.

In later experiments Joule produced heat in water by stirring it with large paddles. Since he neither was a university graduate nor held a recognized scientific appointment, Joule's work at first met with a chilly reception by the scientific community. Papers he submitted to the Royal Society were rejected. Opinions began to change, however, in 1847, after he presented a paper at the meeting in Oxford of the British Association for the Advancement of Science. There he had the good fortune to meet William Thomson, who soon gave Joule strong support.

Thomson (1824–1907; see fig. 35), of Scottish descent, was born in Belfast, Ireland. He was educated at the University of Glasgow and at Peterhouse, Cambridge. He graduated with high honors in 1845, and when he first met Joule in 1847 he had been appointed the year before, at the early age of twenty-two, professor of natural philosophy at the University of Glasgow; he was to hold that position for over half a century. He is best known today by his title of Lord Kelvin, and to avoid confusion with other scientists named Thomson it seems better for us to call him Kelvin from now on, even though he did not receive that title until late in his life.

Kelvin's work covered a wide range, and he made many contributions of the greatest importance to both science and technology. In his earlier years he was intimately concerned with the new science of thermodynamics, and it was he who first used the word "thermo-dynamic," in 1849.

Kelvin was impressed by Joule's paper at the BAAS meeting and afterward had some private discussions with him. Two weeks later Kelvin was on a walking tour in Switzerland and unexpectedly ran into Joule, who was carrying a large thermometer; although on his

**Figure 35.** William Thomson (1824–1907), who later became Lord Kelvin of Largs and is now usually remembered by that name, did work of great significance leading to the second law of thermodynamics and made many other contributions to science. He became a revered national figure and is buried in Westminster Abbey beside Isaac Newton. (Edgar Fahs Smith Collection, University of Pennsylvania Library)

honeymoon, the enthusiastic Joule was making temperature measurements at the top and bottom of a large waterfall while his bride waited patiently in a carriage not far away. These meetings, and Joule's papers, finally convinced Kelvin that Joule was right. Heat must be a mode of motion, not a substance, and in a steam engine there is an actual conversion of heat into mechanical work. It was probably Kelvin who suggested that from the principle of conservation of energy we can formulate the *first law of thermodynamics*. This is a more specific formulation of the principle: it states that the change in the total energy of a system is equal to the work done on it plus the heat supplied to it. In using this relationship the same unit must be used for energy, heat, and work. Joule has the honor of having the official (SI) unit of energy, work, and heat named after him.

▲▲▲

The *second law of thermodynamics* was also arrived at rather gradually and resulted from the realization that the application of the first law to steam engines had some curious features. We have seen that Kelvin was slow to accept the first law, even though he was on friendly terms with Joule and followed his work closely. Paradoxically, the reason for his initial reluctance was that he had studied heat conversion even more deeply than Joule had. Kelvin knew that if there was an interconversion of heat and work, there was something rather odd about it. Work could be converted into heat without any apparent complications, as in Rumford's and Joule's experiments, but there were apparently some restrictions on the conversion of heat into work. Kelvin grappled with this problem for some time and was finally led to a deeper understanding of the restrictions.

It is natural to suppose from the numbering that the second law of thermodynamics would have been discovered after the first law. Instead, the two laws grew up together, and it is only for clarity and convenience that we have so far focused on the first law. Some significant work was being done that led to formulations of the second law at just about the same time that the first law was being generally accepted.

This work was carried out by a remarkable young French military engineer, Nicolas-Léonard-Sadi Carnot (1796–1832; see fig. 36). Carnot was born in Paris and was educated at the École polytechnique as a military engineer. For a few years he held various routine military positions but obtained a protracted leave of absence to undertake study and research in science and engineering. In 1824, when he was twenty-eight years of age, he published a 118-page book in which he developed a highly original treatment of heat engines on the basis of his belief in Lavoisier's *calorique* theory.[2]

**Figure 36.** Nicolas-Léonard-Sadi Carnot (1796–1832) was famous for his analysis of the functioning of the steam engine, a study that led to the formulation of the second law of thermodynamics. (From a painting by Louis-Léopold Boilly [1761–1845], done when Carnot was aged seventeen and wore the uniform of the École Polytechnique in Paris. Edgar Fahs Smith Collection, University of Pennsylvania Library)

Believing that heat is an imponderable fluid, he thought that when heat flows from a higher to a lower temperature and work is done, the heat is actually conserved. Carnot discussed the analogy of a waterfall causing a wheel to turn (refer to fig. 31); there is no loss of water, the wheel being turned by the force of the water. In the same way he thought that the force of falling heat would cause the piston in an engine to move and perform work. Although this idea is wrong, it luckily did not affect his final conclusions.

Carnot considered what was called the maximum "duty" of an engine, which is the amount of work it does when a given amount of fuel is consumed. He expressed the work done as the mass of water that could be lifted multiplied by how high it is lifted. One important conclusion he reached is that if the engine operates between a higher temperature ($T_h$) and a lower temperature ($T_c$), the duty is larger if the difference between the two temperatures ($T_h - T_c$) is larger, just as with a waterfall the greater the distance of fall, the more work is done by the water wheel. For example, if the heat falls from 100 °C to 0 °C, the work done is more than if it falls from 100 °C to 99 °C. Carnot also found that for a given drop in temperature, $T_h - T_c$, the work is greater, the smaller $T_h$ is; thus a drop from 1 °C to 0 °C will produce more work than a drop from 100 °C to 99 °C.

Carnot discussed his conclusions with reference to some of the steam engines of his time. He was able to explain why a high-pressure steam engine is more efficient than a low-pressure one: "It is easy to see why the so-called high-pressure steam engines are better than the low-pressure ones; their advantage lies essentially in their ability to utilize a greater fall of caloric. Steam generated at a higher pressure is also at a higher temperature, and as the temperature of the condenser is nearly always the same, the fall of caloric is evidently higher."[3]

Carnot's work had great practical implications, which he pointed out. Previously it had been thought that engines could be improved by changing to different materials without changing the temperature at which they work. Carnot showed, however, that attention must be directed to the working temperature and not to the materials used.

It was not for nearly twenty years after Carnot's death that the importance of his work began to be appreciated by the scientific world. One of the first to develop Carnot's ideas was Kelvin. We saw earlier that Kelvin was at first skeptical about Joule's conclusions but after a short time accepted them enthusiastically. It was from his study of Carnot's work that Kelvin realized that when an engine operates, all of the heat absorbed cannot be converted into mechanical work. Some heat must be wasted by simply passing from a higher temperature to a lower temperature; Kelvin referred to this as the *dissipation of energy*. He saw that as a result, an engine cannot operate if everything is at the same temperature. For example, a ship cannot propel itself by abstracting heat from the surrounding water; the heat must be obtained from something at a higher temperature, and there must be dissipation of some of the heat, which simply passes from a higher to a lower temperature and does no work.

A machine that, though consistent with the first law, is supposed to violate the second law of thermodynamics is referred to as a *perpetual motion machine of the second kind*. An example would be a ship that operated by withdrawing heat from the surrounding waters. One such ship was, in fact, designed with this idea, by inventor John Ericsson (1803–89), who was born in Sweden but settled in the United States. He built a vessel, named the *Ericsson*, in which heat was supposed to be "regenerated" and used over again. We are not surprised today to learn that it was a disaster; it completely failed its trials in 1853 and had to be refitted with steam engines.

In 1848 Kelvin suggested that Carnot's ideas could be understood more clearly in terms of a specially defined temperature, called the *absolute temperature*. The temperature systems to which most of the world is accustomed today are the nearly identical centigrade and Celsius scales, the main exception being the United States, which still adheres conservatively to the old Fahrenheit scale. Kelvin based his scale on measurements on the work done by an engine when a gas expands reversibly. The theory of gas expansion showed that the work is proportional to a temperature defined as the Celsius temperature plus a constant temperature that we now take to be 273.15 degrees. This temperature is now called the *Kelvin temperature* or the absolute temperature and, in honor of Lord Kelvin, the symbol used is K. The temperature 25.0°C, for example, is 25.0 + 273.15 = 298.15 K. (By convention we do not put the degree sign before the *K*.)

In 1851 Kelvin stated that "it is impossible . . . to derive mechanical effect from any portion of matter by cooling it below the temperature of the coldest of the surrounding objects." This is one way of stating what has come to be called the second law of thermodynamics.

A different way of expressing the second law was suggested by German physicist Rudolf Julian Emmanuel Clausius (1822–88; see fig. 37). Clausius was born in Köslin, then a town in Prussia, but now in Poland and called Koszalin. He was educated at the universities of Berlin and Halle, obtaining his doctorate at the latter in 1847. In 1855 he became professor of mathematical physics at the Polytechnicum in Zurich, moving to the University of Würzburg in 1867 and the University of Bonn in 1869. He remained in Bonn to the end of his life, serving in his later years as rector of the university (or, as we would say in most English-speaking countries, its president or vice chancellor).

Clausius was not an experimentalist but instead worked mainly on theories of molecular mechanics and thermodynamics. In the 1850s and 1860s he carried out research of great significance in thermodynamics, his main contribution being to make a detailed mathematical analysis of Carnot's ideas about steam engines. In the course of this work he arrived at a new physical property, which in 1865 he called *entropy*. Since an exact understanding of entropy involves rather subtle arguments and also the use of integral calculus, we cannot go into it deeply. We can, however, get an idea of entropy by saying that the change in entropy when any process takes place can be calculated by imagining it to occur infinitely slowly

**Figure 37.** German physicist Rudolf Julius Emmanuel Clausius (1822–88) was distinguished for his work on the second law of thermodynamics, in particular for his introduction of the quantity known as entropy. (Edgar Fahs Smith Collection, University of Pennsylvania Library)

under certain ideal conditions; the change in the entropy is then the sum of the infinitesimal amounts of heat absorbed at every stage of this ideal process divided by the absolute temperature at that stage. We recall that Carnot, and later Kelvin, pointed out that when a steam engine operates, some of the heat is unavailable for conversion into work. What Clausius did by introducing entropy is to express this precisely; the entropy change is a measure of the *extent to which heat energy is unavailable for conversion into mechanical work.*

Entropy is so subtle a property that many scientists were unable to understand it when Clausius first suggested it. Kelvin, for example, never appreciated entropy, and maintained that the second law is best understood in terms of the dissipation of heat, which is easily visualized. Kelvin's philosophy of science was that everything must be explained in terms of a mechanical model, but entropy cannot be explained in this way. Properties such as volume, pressure, and temperature can be measured with simple instruments and can be appreciated even by people who do not know much about science. Entropy, on the other hand, is elusive; no instrument can directly measure an entropy change, which has to be calculated in a rather complicated way from data involving heat and temperature changes—to make things worse, they must be hypothetical processes, occurring infinitely slowly.

Clausius concluded that when any spontaneous process occurs, such as a building falling down or an explosion occurring, there must be an increase in the entropy of the universe as a whole. The increase need not be in the system itself; there can be a loss of entropy in the system, and the process will occur if there is a greater gain in the surroundings, resulting from the emission of heat by the system. For example, when hydrogen and oxygen gases combine together to form water, there is a loss of entropy in the molecular system itself. A large amount of heat is, however, given out, which causes a large increase in the entropy of the surroundings, more than enough to cover the loss of entropy in the system.

In his 1865 paper Clausius expressed the two laws of thermodynamics in a compact form, as follows:

*The energy of the universe remains constant.*
*The entropy of the universe tends toward a maximum.*

These two laws are sometimes expressed in everyday language as

*You can't get something for nothing.*
*You can't even break even.*

▲▲▲

We should now discuss in more detail just what entropy is and why the entropy of the universe constantly increases, as the second law of thermodynamics (in Clausius's version) says it must. This turned out to be a rather difficult question to answer, and it took several decades, until well into the twentieth century, for there to be general agreement among the experts.

The two main answers to the question may be stated briefly as follows. The first, the correct one, is that it is all a matter of probability. There is a natural tendency for systems to pass from orderly states to states of greater disorder, just because the disordered states are more probable. The example of a deck of cards is often used to explain entropy. A deck can be arranged in a particular way, such as the way the cards are arranged by the manufacturer in an unopened pack. We will arbitrarily call that, and that only, an *ordered deck*. We can shuffle the deck, and what we call a *shuffled* or *disordered deck* is obviously much more probable than the ordered one. This is because there is an enormously large number of arrangements that we call disordered, whereas only one qualifies as ordered. Shuffling an ordered deck will almost certainly produce a disordered deck, and it is highly unlikely that a shuffled deck will become ordered if it is further shuffled. According to this view the second law, unlike the first one, is not an absolute law: it is possible for a system to go from a disordered state to a more ordered one, just as a shuffled deck of cards may, upon further shuffling, become an ordered one. Such things are not impossible; they are just extremely unlikely. We can easily calculate that if a person shuffled a deck of cards at the rate of one shuffle per second, even after a period of time equal to the age of the universe (about 12 billion years) it is extremely unlikely that a specified ordered deck would have been obtained (discussed later in greater detail).

The second answer to the question, the one that proved to be wrong, is that nothing more is required than the laws of mechanics. The view taken was that from these laws one could arrive, purely mathematically, at the conclusion that a process such as the mixing of two gases would inevitably occur in the direction of mixing and could never move in the opposite direction. On this view a violation of the second law would be absolutely impossible, not just highly unlikely.

The first answer, the correct one, may be expanded as follows. We are all familiar with processes that occur naturally as a result of this tendency for an ordered state to become a disordered one. When a lump of sugar is dissolved in coffee we know that the molecules of the sugar spread themselves throughout the liquid; however long we wait, we do not see the cube reforming itself—although if we could wait a very long time (almost certainly longer than the age of the universe!) it might do so, but it would dissolve again at once. We know that if a bottle of perfume is left open, the perfume will spread around the room, and we do not expect that in our lifetime that the molecules of perfume will go back again into the bottle. Some of us have seen a demonstration experiment in which oxygen and hydrogen gases are brought together, and a flame is put to the mixture; the gases explode with the formation of water ($H_2O$). However, the opposite process does not occur. However long we wait, a glass of water will not suddenly decompose into hydrogen and oxygen. The reason is that there is a great increase of entropy, in both the gases and the environment, when the gases are exploded together—as we saw, largely because heat is given off and is dissipated into the surroundings. In principle, heat from the surroundings could assemble in a glass of water and decompose it into hydrogen and oxygen, but the probability of this happening is exceedingly remote.

Note that time enters into this argument; disorder increases as time passes. British astronomer Sir Arthur Eddington (1882–1944) looked at the matter from this point of view and referred to entropy as the "arrow of time." At a later time the state of the universe has a greater probability than at an earlier time, and as a result time cannot go backward but only forward.

An instructive way of thinking about the concept of entropy was suggested in 1867 by James Clerk Maxwell, whose work on light and the electromagnetic theory we encountered in the last chapter. Maxwell's idea depended on the existence of an imaginary creature that came to be called "Maxwell's demon."[4] This creature was born in a letter that Maxwell wrote on December 11, 1867, to an old friend, Peter Guthrie Tait. The point of this letter was to show how, in principle (but almost never in practice), the second law could be violated.

Maxwell considered a vessel divided into two compartments, A and B, separated by a partition with a hole in it that could be opened or closed by "a slide without mass" (see fig. 38). The gas in A was at a higher temperature than the gas in B. In a gas at a higher temperature, the average speed of the molecules is greater than if the gas is cooler, but there is a distribution of speed. Maxwell imagined "a finite being," later called a demon, who knew the speeds of all the molecules. This creature could open the hole for an approaching molecule in A when its speed was low and would allow a molecule from B to pass through the hole into A only when its speed was high. As a result of this process, said Maxwell, "the hot system has got hotter and the cold colder and yet no work has been done."

Of course, Maxwell did not imagine that his "finite being" existed; he emphasized that his intention in inventing it had simply been to provide us with an understanding of why the second law of thermodynamics applies; it is just a matter of probability. In spite of clear statements by Maxwell, his demon has been widely misunderstood. Maxwell made it clear that he regarded his demon as no more than a device to help people to understand the second law. He made this clear, for example, in a letter to J. W. Strutt (later Lord Rayleigh), written in December 1870: "The 2nd law of thermodynamics has the same degree of truth as the statement that if you throw a tumblerful of water into the sea, you cannot get the same tumblerful out again." He meant, of course, exactly the same molecules as before.

It is interesting that most of the earlier discussions of the second

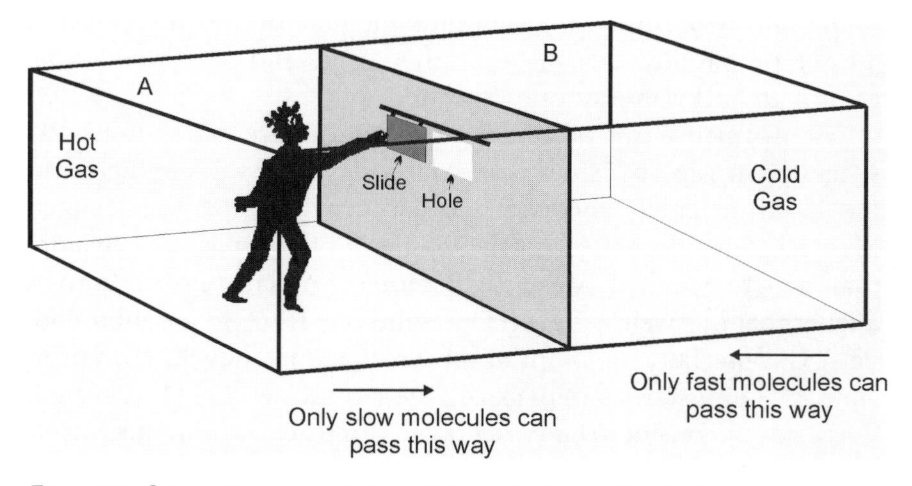


**Figure 38.** Schematic representation of Maxwell's demon. The left-hand compartment, *A*, contains gas that is hotter than the gas in *B*, which means that the molecules in *A* are on the average moving faster than the molecules in *B*. The demon allows only fast-moving molecules to move from right to left and only slow molecules to move from left to right. The hot gas therefore gets hotter, and the cold gas cooler! Maxwell used this imaginary device to explain the second law of thermodynamics.

law made no reference to a process that today is often used as a good example of the application of the law, namely the mixing of two different gases at constant temperature. Maxwell's demon could reverse such a process; air, for example, could be separated into oxygen and nitrogen by the use of a demon who would allow only oxygen molecules to pass in one direction and only nitrogen molecules to pass in the other. The earlier workers thought of the second law as relating only to engines, where there is a passage of heat from a higher temperature to a lower one, and a mixing process is not obviously related to such transfers of heat.

One particular point about entropy and the second law requires mention. It relates to the role of information in connection with the law and what has come to be called *Gibbs's paradox,* named after eminent American applied mathematician Josiah Willard Gibbs (1839–1903), who made important contributions to thermodynamics. Imagine a vessel divided into two halves separated by a partition. Suppose first that the two compartments contain two different gases at the same pressure and temperature. If the partition is then removed, there will be an increase of entropy because the gases have mixed, and the value of the entropy is easily calculated. The increase in entropy is due to the increase in randomness or disorder.

Gibbs compared this with what happens if the two gases, instead of being different, are identical. There is then obviously no increase of entropy when the partition is removed, because nothing in effect has happened; there is no longer any mixing of gases. This difference sometimes causes consternation, because from one point of view what has happened in the two cases is essentially the same. Gibbs himself did not regard the difference in the two entropy changes as really paradoxical. He saw clearly that the difference depends on the information we have about whether the gases are the same or not. Obviously, in computing an entropy change, we have to take into account the relevant information.

The same point arises if we are considering the entropy change in the shuffling of fifty-two playing cards. We can say nothing about the entropy change unless we have further information. If by mistake the cards have been printed so that all are identical, there will be no entropy change on shuffling. If, however, the cards have been printed normally, there is an increase in entropy when an ordered deck is shuffled. We call this *informational entropy* or *conformational entropy*, and it arises from the fact that for an ordered deck we accept only one arrangement of cards out of the vast number of possible arrangements. Probability theory leads to the result that the possible number of arrangements in a deck of fifty-two cards is about $4.4 \times 10^{66}$. To get an idea of the enormity of this number, suppose that we have a shuffled deck and that we shuffle it at the rate of one shuffle a second, how long would it be before we would have a sporting chance of getting the ordered arrangement? The answer is obviously about $4.4 \times 10^{66}$ seconds. About 12 billion years have elapsed since the beginning of time, and that is equal to only (!) about $10^{17}$ seconds. To get the ordered deck we would probably have to shuffle for about $10^{51}$ times as long as that.

When we mix two gases the informational entropy is obviously much larger than for the shuffling of cards, because the number of molecules in a gas of visible size is much greater than the number of cards. Suppose for example that we have a liter (roughly a quart) of ordinary water mixed with a liter of heavy water, $D_2O$. The chance that it will spontaneously separate into the two forms of water works out to be ten raised to the power of $10^{23}$, that is, one followed by $10^{23}$ zeros. This string of zeros would stretch to the Moon and back about 100 billion times.

This matter of informational contributions to entropy is of great importance but is widely misunderstood. A proper understanding of it has formed the foundation of modern information theory. Sometimes the fundamentals of this subject are presented with reference to the activities of Maxwell's demon. If a demon did reverse the ordinary course of events and cause a warm gas to get warmer and a cold one to get colder when the two gases are side by side (as in fig. 38), it would be by acquiring information about the speeds of the individual molecules. If he did so, and the gases did incur a decrease in entropy, this would be overcome by the entropy change that is involved in the use of information.

▲▲▲

Important contributions to understanding the second law of thermodynamics were also made by Austrian physicist Ludwig Boltzmann (1844–1906; see fig. 39). Boltzmann's work is also consistent with the conclusion that the second law is all a matter of chance or probability. Unfortunately he somewhat clouded the issue by suggesting at times that the law could be derived without taking chance into account.

Boltzmann was born in Vienna and attended the University of Vienna. He received his doctorate in 1866 and continued there for a year as a research assistant. Boltzmann had a pathologically unhappy and discontented disposition and made a number of moves during his career, being at various times at the universities of Vienna, Graz, Munich, and Leipzig. At the end of his career, in 1902, he returned to Vienna. During his later years there he suffered a mental breakdown, becoming incapable of lecturing or carrying out any of his academic duties. He made several attempts to take his own life,

**Figure 39.** Austrian physicist Ludwig Boltzmann (1844–1906) developed a kinetic theory of the approach to equilibrium; his work led to the important new branch of science called statistical mechanics.

spent some time as a voluntary patient in a mental hospital, and finally committed suicide.

We noted earlier in this chapter that an alternative but incorrect explanation of the second law of thermodynamics was that it can be derived from the kinetic theory of gases and that probability was not involved. As a young student, before he had studied Maxwell's papers, Boltzmann held the mistaken opinion that this was the correct explanation, and in 1866 he wrote a paper on the subject. Shortly afterward Clausius wrote a paper along the same lines, which was at once followed by one from Boltzmann, pointing out that his 1866 paper included essentially the same treatment as that given by Clausius and concluding with the rather sardonic comment: "I can only express pleasure that an authority with Herr Clausius's reputation is helping to spread the knowledge of my work in thermodynamics." This comment, which is typical of Boltzmann's unsophisticated and clumsy style, probably irritated Clausius, already a distinguished professor. However, in a paper published in 1872 Clausius graciously conceded Boltzmann's claim, explaining that extraordinary demands on his time had made it difficult for him to follow the scientific literature.

While these exchanges were taking place, Maxwell was regarding them with some amusement, realizing that both Clausius and Boltzmann had missed the point. Boltzmann later studied Maxwell's papers more carefully and then realized that it was wrong to ignore probability; his later papers developed the probability arguments further.

In 1868, two years after his unsuccessful attempt to explain the second law without the use of probability, Boltzmann published a paper of great importance. Maxwell had previously derived an expression for the fraction of molecules in a system that have a particular speed. This is known as the *Maxwell distribution law*, another important contribution for which he is famous. What Boltzmann did was to extend Maxwell's theory and to give an expression for the fraction of molecules present in a system that have a particular energy, $E$. He found that this fraction is proportional to the fraction $\exp(-E/k_BT)$, where $k_B$ is now called the *Boltzmann constant* and $T$ is the absolute (Kelvin) temperature. This formula requires some explanation. The expresssion $\exp(-E/k_BT)$ is just another way of writing the exponential $e^{-E/k_BT}$. By this we mean that the number $e$ is raised to the power of $-E/k_BT$. This number $e$, which has the value of about 2.71828, turns up often in scientific work. When we write the exponential $e^{-E/k_BT}$ we mean that the number $e$ is multiplied by itself $-E/k_BT$ times, just as when we write $10^2$ we mean that 10 is multiplied by itself twice. Remember that when we write 10 raised to a negative power, such as $10^{-3}$, the result is a fraction, in this case 0.001, or one-thousandth. Thus, since $-E/k_BT$ is a negative quantity, $\exp(-E/k_BT)$ is a fraction. The great importance of this derivation by Boltzmann is that he had shown that the fraction $\exp(-E/k_BT)$ not only applies to the energy of a molecule that is due to its motion from one place to another (its kinetic energy of translation, as we call it); it applies to all kinds of energy, such as energy due to a gravitational field and internal energy of vibration. One feature of Boltzmann's derivation is that he regarded the total amount of energy as distributed among molecules in such a manner that all ways of doing so are equally probable.

An important consequence of the Boltzmann factor, exp $(-E/k_BT)$, is that it leads to the useful conclusion that the average energy of a molecule at temperature $T$ will be something like $k_BT$, depending to some extent on the nature of the molecule.

Boltzmann made these major contributions, and a few others, while still not yet twenty-five; by that time he had published a total of eight papers, and already his work was beginning to attract attention. From 1871 to 1887 he published several papers of great significance in which he showed how his factor allowed all of the properties of a system to be calculated. This is what the field of *statistical mechanics* is mainly concerned with, and the methods Boltzmann used are essentially those commonly used today. What he did in essence was to consider the distribution of energy in terms of putting atoms into pigeonholes corresponding to different values of their energy. The total amount of energy has to be the same for each arrangement, but this condition can be satisfied in a vast number of different ways. Arguing in this way, he proceeded to show how to calculate the number of equivalent ways in which the atoms could be fitted into the compartments (subject to the condition that the total energy is fixed). This number he equated to the probability of the atomic distributions. He showed that the most likely distribution was the one given by the distribution equation.

He went further by pointing out that the closer that any distribution was to that of equilibrium, the more probable it was. By reasoning in this way Boltzmann was led to his famous relationship between entropy and probability, $W$:

$$S = k_B \log W.$$

(See "A Few Points on Mathematics and Measurement" for an explanation of logarithms. The logarithm used here is not the common logarithm but the natural logarithm, now usually written as ln $W$, which is the power to which the number $e$ has to be raised to get the number in question.) In this equation, which is engraved on Boltzmann's tombstone in Vienna, $W$ is the number of possible ways of making a given distribution of atoms or molecules, corresponding to a given total energy of the system. These molecular configurations are now referred to as *microstates*. This equation allows an expression for entropy to be obtained from the statistical distribution, and from this expression the other thermodynamic properties can be calculated.

With this relationship Boltzmann had actually introduced an extension to the Clausius definition of entropy. We saw earlier in this chapter, with reference to Gibbs's paradox, that when considering an entropy change it is important to include the informational entropy contributions. For example, in dealing with the entropy change when two gases are brought together, we get a different answer using the information that the two gases are different from what we get if we think they are identical. If we proceed by using Clausius's definition we have to add the informational entropy rather arbitrarily. The Boltzmann formulation, on the other hand, is more satisfactory in that it deals with the informational contribution quite naturally.

Work on the second law of thermodynamics raised a significant point about the laws of nature: every previous law of nature that had been deduced was quite definite and straightforward. The second law was the only law that had a different basis; it was not *absolutely* true but was only approximately true, as a matter of probability. It was true that the chance of the law being violated was usually exceedingly remote. Some scientists, however, found it philosophically unsatisfactory that what was supposed to be a law of nature was true only by chance.

In other words, the second law presented the first important challenge to the purely deterministic laws of nature—to the idea that every cause has a unique result. In later chapters we will come back to this concept and will see that evidence of an entirely different nature leads to the same conclusion. From the present state of the universe we cannot predict precisely what the future will be. For several reasons there must always be uncertainty.

▲▲▲

What has been discussed in the present chapter is called classical physics, which means that it is somewhat old-fashioned. That is not to say that it is wrong; recent work has greatly supplemented it but has not disproved any of it. A scientific writer a hundred or so years ago might have covered much the same material as included in this chapter, but there would have been one important difference: these final comments would not have been made, because the writer would then have thought that energy was completely understood. For example, eminent physicist Lord Kelvin wrote much about energy, and at the end of the nineteenth century he was apt to say that not much remained to be discovered about it. Indeed, at the beginning of the twentieth century most scientists felt satisfied that all their important problems were more or less solved. The principles of mechanics had been well worked out, heat was understood to be a mode of motion, and light to have wave properties. Some exciting discoveries had been made toward the end of the nineteenth century—radio transmission, the electron, x-rays, and radioactivity—and to many scientists that seemed to be all that could ever be discovered. A few problems remained, but it appeared that all that was needed was to fill in some details.

How wrong they were! Little did anyone guess that within the first five years of the new century, two new theories would bring about radical changes in the way we think about science, and that they were to have great practical consequences. These two theories were the quantum theory, introduced by Max Planck in 1900 and developed by Albert Einstein in 1905, and the theory of relativity, due entirely to Einstein in 1905 and later. Both of these had implications for the understanding of energy, and we will talk about them in the next two chapters. We will first see that energy can no longer be assumed to be continuous; it comes in tiny packages that are too small to recognize by ordinary methods of observation and can only be detected by special techniques. Although these packages are tiny and hard to detect, they brought about fundamental changes in our understanding of the universe.

4

# PACKETS OF ENERGY

So far we have thought of energy as if it were continuous, by which we mean that we can have any amount of it we like. Until 1900 almost everyone assumed that this was the case, since nothing known up to that time seemed to suggest otherwise. Today we can think of several things that were known before 1900 that can only be properly explained if energy comes in packets rather than being continuous. For example, it was known that the spectra of substances exhibit lines, or bands of lines, at particular wavelengths. Why only at these wavelengths? We know now that it is because energy can only exist in particular amounts. Until the end of the nineteenth century, though, few scientists gave much thought to this question; Maxwell did so but could not think of any answer. The time was not ripe for any such explanation.

These packets are called *quanta*, from the Latin *quantum*, "how much?" and the theory that is concerned with this theory is called the *quantum theory*. This theory does not intrude itself in an obvious way into our everyday lives. For example, when we drive a car only certain speeds are allowed to us, but it is impossible for us to know this. The reason for our ignorance is that the permitted speeds are so fantastically close together that the most careful mechanical measurements would never detect that there is any quantization. An analogy is provided by the money we use, which is quantized, but in most currencies the quanta are so small that we are not bothered by them. Whether we have 29.42 or 29.43 (dollars, Euros, or whatever) in our pockets really doesn't much matter. The quanta imposed by distributors of milk, on the other hand, are larger and more noticeable. If we are eccentric enough to try to buy 1.63 pints (or liters, or whatever) of milk we will be refused; today milk usually is sold in quanta of half a pint, half a liter, half a gallon, etc.

The quanta that exist in nature are always exceedingly small, much smaller than those that we humans impose on ourselves with such things as coins and milk. For one thing, quanta are on the atomic scale, so when we are concerned only with the things that we see around us they are undetectable. Journalists, politicians, and others are fond of talking about "quantum leaps" apparently without realizing that in everyday life these "leaps" are quite undetectable. Whenever you hear a politician promising a quantum leap, meaning that it is big and important, remember that in reality, changes from one quantum state to another are always tiny and usually occur in a completely unpredictable way!

▲▲▲

Before we can understand the quantum theory we need to know something about vibrators, or oscillators. Imagine a weight (strictly speaking, a mass) suspended at the end of a spring. We will start with it motionless, as shown in figure 40, *a*. If we pull it down to position A, as shown in *b*, and then release it, the weight will then vibrate up and down between positions A and B. The distance between A and B is known as the *amplitude* of the vibrations. Another important characteristic of the vibration is the time that it takes for the weight to move from position A up to B and back again to A; this time is called the *period* of the vibration. The number 1 divided by the period (a value that we call the *reciprocal* of the period) is called the *frequency* of the vibration, usually given the symbol $\nu$ (Greek *nu*). Suppose, for example, that it took two seconds for the weight to move from A up to B and back to A; we would say that the period was two seconds and that the frequency, $\nu$, was 1

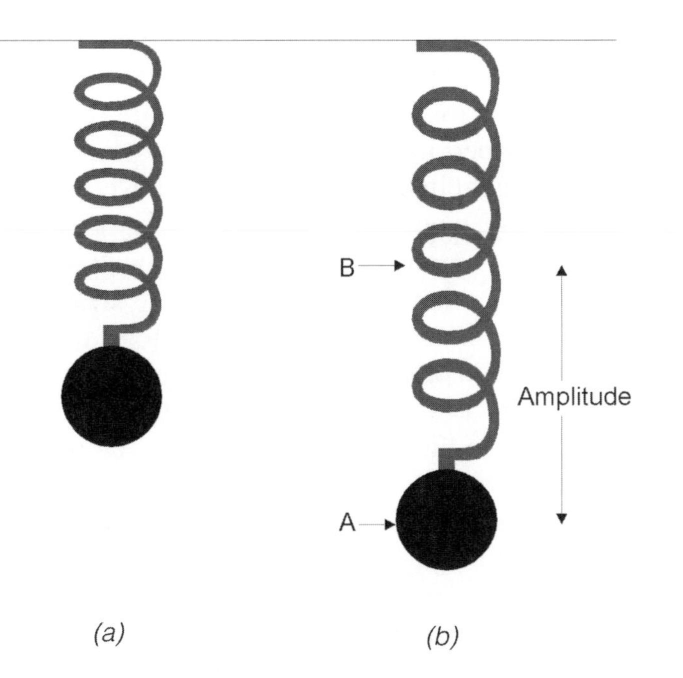


**Figure 40.** *a*, A weight attached to a spring, in its undisturbed state; *b*, the weight pulled down to position *A*. When it is released the spring pulls it back to position *B*, after which it oscillates between the two positions. By definition the amplitude of the vibration is the distance *AB*. The time that it takes to travel from *A* to *B* and back is called the period of the vibration, and the reciprocal of that time is the frequency of the vibration.

divided by 2, or 0.5 per second, which we can call 0.5 reciprocal seconds or 0.5 hertz (0.5 Hz).

Similar laws of motion arise in other oscillating systems that we meet in scientific work. A pendulum, such as that of a clock, or a child's swing, behaves in the same way. It also undergoes a to-and-fro motion that obeys much the same laws as the weight hanging from the spring. The atoms in a solid also are constantly undergoing a vibrational motion, and they are often referred to as oscillators. Hydrogen gas is composed of molecules that are diatomic, by which we mean that each of the molecules consists of two atoms held together by a chemical bond having the same characteristics as the spring shown in figure 40. We can think of the bond between the two hydrogen atoms as continuously undergoing vibration, in which it alternately shortens and lengthens.

To a good approximation, the frequency remains the same whatever the amplitude or period of the vibration. Suppose, for example, that we pulled the weight in figure 40 down to a lower position than A, which means that we would be using more energy since we would be stretching the spring more. After we released the weight it would rise to a position higher than B, which means that the amplitude of the vibration is larger than it was before. In spite of this, however, the frequency (and therefore the period) will remain almost exactly the same. When there is a greater amplitude, the weight travels a greater distance between the extremities of each vibration. It does so in the same period of time, which means that it has to move faster on the average to make the longer trip, and it therefore has more energy. Vibrations with a greater amplitude are thus more energetic but have much the same frequency.

With a swing or a pendulum we take it for granted that, by adjusting the length of the string, we can make the frequency of oscillation anything we like. The quantum theory says that this is not really the case; only certain energies are possible. However, these quantum restrictions are too small to matter in a clock or a child's swing; for them to be noticeable we must deal with matter at the atomic level.

Thus if we are dealing with an atom vibrating in a solid, or a vibrating hydrogen molecule, we have to recognize that only certain amplitudes of vibration, and therefore only certain energies, are possible. This quantization may be ignored for visible objects like swings, but when we get to the atomic scale, quantum effects have to be taken into account. This is why quantum effects are not noticed by us in our everyday lives. It is also why scientists were unable to find any evidence for the quantum theory until the twentieth century. All their work had been done on much too large a scale.

▲▲▲

The quantum theory had a rather obscure beginning, which explains why, in spite of its great significance, it was not accepted very quickly.

It was first proposed as a way of explaining some results on how the energy of radiation emitted by a hot body varies with the wavelength or frequency of the radiation. Around the turn of the century a number of physicists, particularly in Berlin, carried out careful experiments on this problem, and attempts were being made to explain the results. The first successful explanation was proposed by Max Planck, whose theoretical treatment of the radiation results was found to have much broader implications than he realized at first.

Max Karl Ernst Planck (1858–1947; see fig. 41) was born in Kiel and studied at the universities of Munich and Berlin. He became professor of physics at Berlin in 1889, and much of his early research was in thermodynamics. On October 25, 1900, there was an important scientific meeting in Berlin, at which some new results on radi-

**Figure 41.** Max Planck (1858–1947) was professor of physics at the University of Berlin from 1889. In 1900 he explained the distribution of energy in radiation in terms of the idea that energy comes in small packets, or quanta; for this work he was awarded the Nobel Prize for physics in 1918. (Edgar Fahs Smith Collection, University of Pennsylvania Library)

ation were presented to the Academy of Sciences. At the meeting Planck suggested an empirical equation that seemed to fit the results better than any that had been suggested earlier. (The word *empirical* is meant to indicate that Planck designed the equation just so that it would fit the data. At this stage he had no theoretical reason at all for suggesting it.)

It turned out that Planck's empirical equation fit the data perfectly, and Planck proceeded to give it a theoretical justification. The essence of the treatment he produced was that a solid body consists of an array of atoms, each one of which is constantly vibrating, obeying the same laws as we have been discussing. He considered the distribution of the energy that these oscillators could have and carried out a statistical treatment, making use of the methods that had been worked out by Ludwig Boltzmann and that we mentioned briefly in the last chapter.

Planck followed this procedure and assumed that if the atoms were oscillating with a frequency $\nu$ their energy came in packets of size $h\nu$, where $h$ is a constant now known as the *Planck constant*. In other words, the energy of the oscillation can be 0, $h\nu$, $2h\nu$, $3h\nu$, etc., but nothing in between. He did this as a mathematical convenience, and the usual procedure at the end would have been to make $\nu$ *zero*—in other words, to make the energy packets have zero size. However, Planck did not take this final step, and he did not at first appreciate the great significance of this omission. He presented the paper that gave this treatment to the German Physical Society on December 14, 1900, which is often regarded as the birthday of the quantum theory.

In view of the fact that we now accept Planck's quantum theory as correct and of pioneering importance, we would have expected his 1900 announcement to have attracted immediate attention. Surely there would have been at least some discussion of it, and perhaps criticism. But in fact, for about five years hardly any notice at all was taken of Planck's paper. The main reason for this is that the theory was presented only as applying to the radiation emitted by hot solids, and its wider implications were not recognized until later. Most physicists at the time were working on what appeared to be much more exciting problems than radiation from hot bodies, such as radioactivity, x-rays, and the electron.

When attention was finally paid to Planck's theory, it was at first usually unfavorable. The criticism was not so much that Planck had suggested an unacceptable quantum hypothesis but rather that he had just made a silly mistake in his mathematics—by not giving the energy packets a size of zero!

At first Planck thought that his quantum theory applied only to oscillators—for example, to the atoms in a solid, which could possess energy only in multiples of $h\nu$. The first suggestion that radiation itself was quantized was made in 1905 by Albert Einstein. In 1905 he published three papers of supreme importance (see the next chapter), and the one that actually led to his 1921 Nobel Prize was on the quantization of radiation. His idea was that light of frequency can be regarded as if it were a beam of particles, having no mass but each one having energy $h\nu$.

Modern textbooks often say that Einstein's 1905 paper on the quanta of radiation was based on Planck's 1900 paper on the quantization of oscillator energies, but only someone who had not looked at the paper could write that. In fact, Einstein made only a passing reference to Planck's work, which at the time he did not find convincing, and he made no use of Planck's constant $h$. Instead his proposal that radiation is quantized was based on a number of other considerations, one of which is called the *photoelectric effect*. This refers to the fact that if light of suitable frequency strikes a solid such as a metal, electrons are emitted.

This effect had been discovered in 1887 at the University of Kiel by Hertz in the course of his famous experiments that led to radio transmission, considered earlier. Rather paradoxically, it was a series of his experiments that convincingly confirmed Maxwell's theory of radiation, based on the idea that light involves vibrations; it also produced some of the first evidence for the particle nature of radiation. Two years later, in 1889, detailed studies of the emission of electrons under the influence of light were reported by Johann Elster (1854–1920) and Hans Geitel (1855–1923). These men were often known as the Castor and Pollux of science, or as the "Heavenly

Twins," since they were inseparable both in their private lives and in their teaching and research. They had been friends at school, and both became teachers at a *Gymnasium* in Wolfenbüttel, near Braunschweig. When Elster married and moved into a house, Geitel boarded in it, and the two friends carried out research in a laboratory that they established in the house. They were often confused with one another, and a man who somewhat resembled Geitel was once addressed by a stranger as "Herr Elster." His reply was something like, "You are wrong on two counts: first, I am not Elster but Geitel, and secondly I am not Geitel."

The experiments done on the photoelectric effect revealed a rather remarkable result. If the frequency of the light was less than a certain value $\nu_0$, there was no emission at all, however strong the intensity of the light. Moreover, if electrons were emitted, the energy of each of them was proportional to the difference between the frequency $\nu$ of the light and the critical frequency $\nu_0$. It did not depend at all on the intensity of the light, which merely affected the number of electrons emitted and not their energy. Einstein realized that the results on the photoelectric effect are inexplicable in terms of the wave theory of radiation, according to which radiation of sufficiently high intensity, irrespective of frequency, would be able to eject electrons. With the particle theory, on the other hand, the explanation is straightforward (see fig. 42). A particle of radiation has energy $h\nu$, and is capable of ejecting an electron provided that $h\nu$ is sufficiently large, i.e., that the frequency $\nu$ is sufficiently large. At first these particles of radiation were just called quanta of radiation, which is still sometimes used, but in 1926 American chemist Gilbert Newton Lewis suggested the name *photon*, by which name they are commonly called today.

Einstein's paper is often referred to today as his paper on the photoelectric effect, but much of the paper is concerned with other matters, such as the way in which atoms and molecules absorb and emit radiation. In other words, it was important in leading to explanations of the detailed nature of spectra.

Einstein's suggestion that light can act as a stream of particles seemed rather surprising at the time. As we saw in chapter 2, the

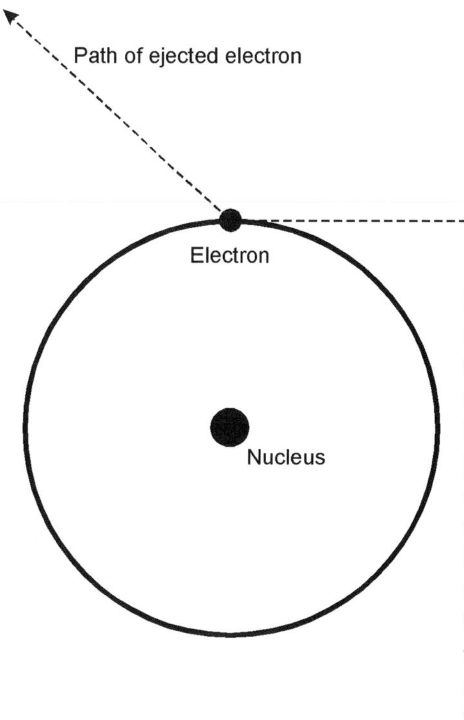


**Figure 42.** Explanation of the photoelectric effect. An electron is shown in its motion around the nucleus, and a quantum of radiation (a photon) interacts with it. The energy of the photon is $h\nu$, where $\nu$ is the frequency of the radiation. A certain minimum energy, called the work function, is required to remove the electron. If the photon does not have this energy, nothing can happen, whatever the intensity of the light. If $h\nu$ is greater than the work function, the photon will remove the electron from the atom and will be annihilated.

wave theory of light became generally accepted in the nineteenth century, and the corpuscular theory seemed only of historical interest. Einstein's suggestion in 1905 that light has particle properties thus seemed incredible, or at least surprising. Also, there appeared to be a logical difficulty with Einstein's theory: the energy of radiation $h\nu$ is expressed in terms of its frequency $\nu$, but frequency only makes sense in terms of the wave theory.

In his 1905 paper Einstein anticipated these objections and answered them. He began his paper by admitting that the wave theory of light was well established and that it certainly applied to properties such as diffraction and interference. But, he continued, properties in which light interacts with matter in bulk relate to *time averages*. However, for some properties, such as the photoelectric effect and the emission and absorption of radiation, there are *one-to-one effects* between the light and the material, and then the wave theory does not apply. In other words, what is needed is a *dual* theory of radiation. Another way of explaining this is to say that light travels as a wave but that when it arrives at its destination it may behave like a particle. For diffraction and interference the wave properties are relevant; for the photoelectric effect and for emission and absorption one must regard the radiation as behaving like a stream of particles. Einstein's suggestion that under certain circumstances light can exhibit particle properties received strong experimental support later, in 1923, when American physicist Arthur Holly Compton (1892–1962) observed the scattering of x-rays.

At first Einstein was not convinced that oscillator energies are quantized and thought that it is only radiation that is quantized. Later he became convinced that Planck was quite right: oscillator energies must also be quantized. He became even more convinced of this when, in 1907, he worked on the specific heats of solids, which are a measure of the heat required to raise the temperature of a solid by one degree Celsius.

Even Planck had difficulty in accepting Einstein's theory of the quantization of radiation—and even in accepting that his own work required us to think about energy in quite a different way. To Planck at first the idea of quanta of energy was no more than an ad hoc hypothesis to explain black-body radiation; he was slow to realize that the quantum theory has a much wider significance. It was in his nature to be conservative, and having been brought up on classical physics he found it difficult to renounce the old ideas. Einstein deserves credit for not only introducing the idea of the quantization of radiation but also for persuading Planck and others that his theory was correct and important.

As time went on, scientists came to realize that the quantum theory had wide applications and that it explained many results that had been impossible to explain without it. There was, for example, a difficulty with the specific heats of solids, which became small as the temperature was reduced, a fact that had seemed inexplicable. In 1907 Einstein showed, however, that if the quantization of the energy of vibration of the atoms of the solid is taken into account, this effect can be explained, and he developed a detailed theory that fit the data perfectly. Then, in 1911, he worked out a similar theory that explained the specific heats of various types of gases. It had earlier been pointed out, particularly by Maxwell, that the specific heats of gases showed some inexplicable behavior. Einstein showed that these difficulties arose from the fact that energy had been regarded as continuous and that when the quantization of energy was taken into account, the difficulties vanished.

There was a further interesting development in 1913, when Einstein and his research assistant Otto Stern (1888–1969) suggested for the first time the existence of a residual energy that all oscillators have, even at absolute zero. They called this residual energy the *Nullpunktsenergie*, which is usually rather unsatisfactorily translated as the *zero-point energy*. They deduced from the specific heat data for hydrogen gas that for an oscillator of frequency $\nu$ the zero-point energy would be $\frac{1}{2}h\nu$. The appearance of the $\frac{1}{2}$ in this expression appeared surprising at the time, but it has been confirmed by experiment and later by quantum-mechanical theory.

▲▲▲

The next important event in the history of the quantum theory was its application to the structure and properties of atoms. This was first done successfully in 1913 by Danish physicist Niels Bohr (1885–1962; see fig. 43). Bohr was born in Copenhagen and studied at the University of Copenhagen, where he obtained his doctorate in 1911. He then worked for a short period at Cambridge with J. J. Thomson and then at the University of Manchester with Ernest Rutherford. It was there that he began to formulate a theory of the hydrogen atom and of similar atoms. After his return to Copenhagen in 1913 he completed this work. In his theory he showed that if he took into account the fact that the energy of the electrons in their orbits was quantized, he could for the first time interpret the properties of the atoms, including some of the details of their spectra. This work was of great significance in leading the way to the improved treatment of atoms and molecules based on the new science of quantum mechanics. Bohr was awarded the Nobel Prize for physics in 1922.

**Figure 43.** Niels Bohr (1885–1962) was particularly famous for his theory of the atom, which he developed in 1913 at the University of Manchester, where Ernest Rutherford was professor of physics. Later he became professor at the University of Copenhagen and headed an institute financed by the Carlsberg breweries. He exerted a strong influence on atomic and nuclear physics, not only through his own work but by his support of others. (Edgar Fahs Smith Collection, University of Pennsylvania Library; photo by Lotte Meitner-Graf)

Bohr's theory of the hydrogen atom provided a foundation for understanding the spectra of all atoms. In the nineteenth century it had been impossible to explain spectra. Why should spectral lines appear in odd places in the spectrum? For example, a spectroscopist looking at the spectrum of a sodium flame observes a sharp line in the yellow region. Why is light emitted particularly at this frequency and not at other ones? It is because only certain orbits are allowed by the quantum theory. The sodium atom happens to have two electron orbits differing in energy by an amount that corresponds to this yellow line. This is illustrated in figure 44, which shows two possible orbits for an electron. An electron moving in one orbit can move into another orbit, farther from the atomic nucleus. To do this it must gain energy, which it can do by absorbing a photon of light of such a frequency that $h$ is equal to the energy required. Alternatively, if an electron drops from the outer orbit to the inner one, the atom emits light of the same frequency.

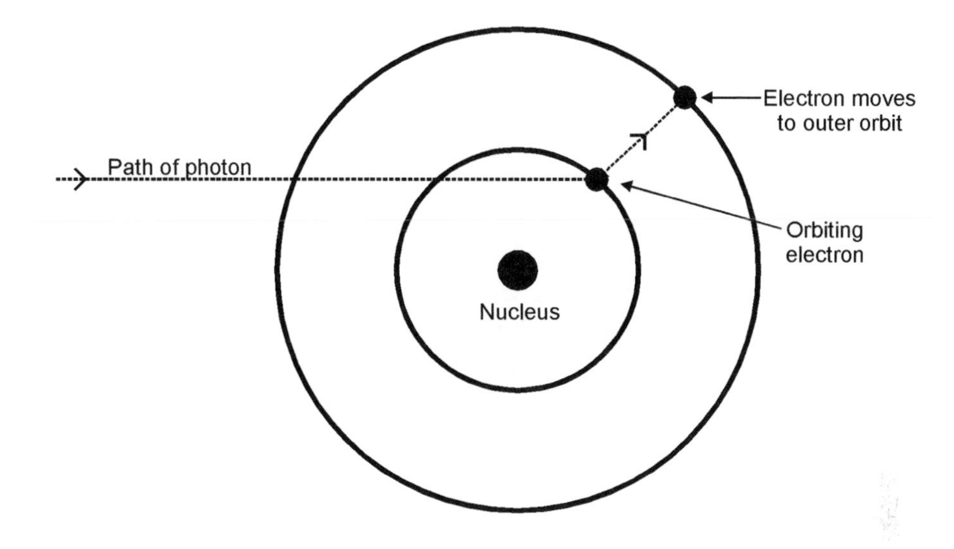


**Figure 44.** The absorption of a photon of light by an orbiting electron. The quantum condition restricts the electron to certain orbits, two of which are shown. The frequency, ν, of the absorbed radiation is determined by the fact that $h\nu$ must be equal to the difference between the energies of the two levels. If, however, the electron is in the outer orbit and drops to the inner one, there is emission of radiation, of the same frequency.

The Bohr theory of the atom was a remarkable achievement, since it provided a general interpretation of atomic structures and spectra, which previously had seemed quite incomprehensible. However, even with the improvements that were made to the theory, serious difficulties remained, as Bohr himself freely admitted. For one thing, it was felt to be arbitrary simply to add quantum restrictions to the old mechanics. Also, it did not seem possible to extend the Bohr theory in such a way as to give a satisfactory treatment of atoms containing more than one electron, or even of the simplest of molecules. What was needed instead was a new mechanics in which the quantum restrictions would emerge as a mathematical necessity and would explain the more complicated molecular systems.

This was accomplished by several physicists in the 1920s, and the result was what we call *quantum mechanics*. The first to do so was Werner Heisenberg (1901–76), who obtained his doctorate in 1923 at the University of Göttingen under the direction of Max Born (1882–1970). He remained in Göttingen doing research with Born, and in the spring of 1925 he developed a treatment of mechanics that made use of an unusual type of mathematics known as matrix mechanics. Born had contributed greatly to the work but suggested that Heisenberg should publish it in his name only. The theory, referred to as quantum mechanics, was soon recognized as being a great advance, and Heisenberg won the 1932 Nobel Prize in physics for his system of quantum mechanics. It was always a matter of regret to him that Born had no share in this honor. The award of the prize jointly to Born and Heisenberg would indeed have been more appropriate: Born had not only initiated the idea of a quantum mechanics; he had also shown the true significance of Heisenberg's obscure mathematics. A less generous man would have delayed the submission of Heisenberg's paper until a joint paper of broader significance could be prepared. Born did receive a Nobel Prize much later, in 1954, for his work on the quantum mechanics of collision processes.

Shortly after developing his theory of wave mechanics Heisenberg made another contribution of great significance. This was his *uncertainty principle*, or principle of indeterminacy. According to it, the position of a particle and its momentum (mass multiplied by velocity) cannot both be measured precisely and simultaneously. Alternatively, we can say that the position and the energy of a particle cannot simultaneously be measured precisely. Heisenberg arrived at the principle by carrying out an imaginary experiment in which a beam of light is used to determine the position and momentum of an electron. The light disturbs the electron, so that if we try to measure the momentum or energy we cannot do so without disturbing the electron; it therefore moves, and as a result its position becomes unknown.

Soon after the appearance of Heisenberg's first paper on quantum mechanics, an equally significant contribution was made by Paul Adrien Maurice Dirac (1902–84). Dirac was trained as an electrical engineer at the University of Bristol and in 1923, being unable to obtain employment, secured a research scholarship to work for his PhD at Cambridge. Late in 1925 he published an alternative formulation of the principles of quantum mechanics, a formulation that was equivalent to Heisenberg's mechanics but more comprehensible and ultimately more useful. Dirac's first paper on quantum mechanics was followed by many others. In 1926 he applied his methods to the hydrogen atom, and in papers that appeared in 1928 he combined his system of quantum mechanics with the theory of relativity. The resulting theory was an extremely powerful one, with far-reaching consequences. For example, Dirac was able to deduce from it that there are two quantum numbers for the spin of the electron, a conclusion that had already been reached on the basis of experiment. We can express this in another way by saying that an electron can spin in only two ways, which are often referred to as clockwise and counterclockwise. Dirac also predicted from his theory the existence of an elementary particle having the mass of the electron but a positive charge. This particle, the *positron*, had not yet been discovered experimentally but was found a few years later, in 1932, by American physicist Carl David Anderson (1905–91).

Dirac was an agreeable and kind man, but had many eccentricities.[1] He wrote an important book, *The Principles of Quantum Mechanics*, which first appeared in 1930 and exerted a wide influence among scientists. In 1933 he was awarded a Nobel Prize for his work. When he heard that he had been selected for the Nobel Prize, he wanted to refuse it, saying that he disliked publicity. Rutherford told him that he would experience more publicity if he refused it, and that convinced him to accept.

Another formulation of quantum mechanics in the end turned out to be equivalent to the treatments of Heisenberg and Dirac, but it involved wave theory; as a result, this formulation of quantum mechanics is called *wave mechanics*. We have seen that Einstein showed that radiation has both particle properties as well as wave properties. The converse idea was put forward in 1923—that particles such as electrons can also have wave properties. This suggestion came from French physicist Louis-Victor, prince de Broglie (1892–1987). He had first intended to become a civil servant and studied history. Later, partly under the influence of his elder brother Maurice, duc de Broglie (1875–1960), who was a distinguished experimental physicist, Louis studied physics and philosophy for his PhD at the Sorbonne.

In his doctoral thesis de Broglie suggested the converse hypothesis as well as an equation that related the wavelength of the wave associated with a particle to the mass of the particle and its speed.

When de Broglie presented his PhD thesis in 1924 he had some trouble with his examiners. One of them, for example, dismissed the ideas as "far-fetched." However, de Broglie had sent a copy of his thesis to Einstein, who returned an endorsement that included the phrase "He has lifted a corner of the great veil." In spite of this, the examiners remained skeptical of the work but somewhat reluctantly awarded the degree. American physicist Robert J. Van de Graaff (1901–67), the inventor of an electrostatic generator that bears his name,[2] was present as a student at the defense of the thesis and some time after World War II remarked, "Never has so much gone over the heads of so many." Presumably he had in mind Winston Churchill's famous remark about the Royal Air Force in the Battle of Britain: "Never in the field of human conflict was so much owed by so many to so few."

Since other prominent physicists were also initially skeptical of his theory, de Broglie did not receive his Nobel Prize until 1929, by which time the theory had been confirmed experimentally. The first experiments to support it involved the *diffraction of electrons*. According to de Broglie's theory, electrons accelerated by a potential of about 100 volts should have wavelengths of a size similar to the interatomic spacing in crystals and would therefore be suitable for diffraction experiments. In January 1927 American physicist Clinton Joseph Davisson (1881–1958) succeeded in demonstrating the diffraction of electrons by a single crystal of nickel. In May of the same year, at the University of Aberdeen, George Paget Thomson (1892–1975) and his research student Andrew Reid observed the diffraction of electrons by thin metal films. Ten years later Davisson and Thomson shared the Nobel Prize in physics for this work. It has been commented that J. J. Thomson was awarded the 1906 Nobel Prize for showing that the electron is a particle, while his son G. P. Thomson obtained a 1937 Nobel Prize for showing that it is a wave. Of course, as we have seen, it is both, and both Nobel Prizes were well merited.

De Broglie's idea came too late to influence the quantum-mechanical treatments of Born, Heisenberg, and Dirac. It did, however, inspire the thinking of Erwin Schrödinger (1887–1961), who developed a wave mechanics that eventually turned out to be mathematically equivalent to the quantum-mechanical theories of Heisenberg and Dirac. His quantum mechanics has been particularly popular with chemists, since it provides a more easily visualized representation of atomic structure, in contrast to the rather formal approaches of Heisenberg and Dirac. Schrödinger, an Austrian by birth and another highly eccentric individual, became professor of physics at the University of Zurich in 1921, and it was there that he did his original work in wave mechanics.

Schrödinger's wave mechanics evolved directly from de Broglie's ideas, which he had first considered to be "rubbish" until he was persuaded otherwise. In the words of physicist Hermann Weyl (1885–1955), Schrödinger obtained his inspiration for wave mechanics while engaged in a "late erotic outburst in his life." His amorous exploits were somewhat remarkable. His unprepossessing appearance, with his thick spectacles, hardly corresponds to the popular idea of a Lothario or Casanova, but such he was. In 1920 he married Annemarie Berthel, and although their relationship was punctuated by many stormy episodes they remained together until his death. Schrödinger had no children by his wife, but he had at least three illegitimate daughters.

His wave mechanics was begun during one of his amorous adventures in late 1925, when he stayed at a holiday resort with a mistress while his wife remained in Zurich, and it was perhaps not only the theory that was conceived at that time. The woman involved in that particular encounter has not been identified, but it would seem that she deserved at least a Nobel award as "best supporting actress."

In Schrödinger's wave mechanics, an electron in an atom is treated as a wave rather than as a particle orbiting round the nucleus, which was Bohr's idea. He did not *derive* his wave equation—that is to say, he did not arrive at it by a formal mathematical proof. Instead he proceeded by analogy with the equations used in Maxwell's electromagnetic theory for wave motion in ordinary radiation. His wave equation is a differential equation involving what is called a *wave-function* and an energy $E$. Quantum restrictions are not introduced arbitrarily but appear as a direct consequence of the wave equation, for which no mathematical solution is possible unless the energy has one of a number of permitted values. The restrictions as to the orbits in which an electron may move are a consequence of the need for a wave to fit into the right amount of space.

At first Heisenberg and Dirac were quite critical of Schrödinger's wave mechanics. Max Born, however, approved of it from the start and in 1926 provided a physical interpretation of the wavefunction. For a problem in atomic or molecular structure, the wavefunction according to Born is related to the probability that an electron is present in a particular small region of space. In many situations the probability is simply represented by the square of the wavefunction.

The misgivings of Heisenberg and Dirac about Schrödinger's theory evaporated when later in 1926 Schrödinger proved that his and the other formulations were mathematically equivalent—they said essentially the same thing but in different mathematical language. Schrödinger showed that each physical property can be replaced by an appropriate mathematical operator and that his wave equation was then obtained. Modern quantum-mechanical calculations are often based on this procedure.

Heisenberg's uncertainty principle and Born's interpretation of the square of the wavefunction as representing a probability are important components of what came to be called the *Copenhagen interpretation* of quantum mechanics, since Niels Bohr, professor at Copenhagen, had much to do with formulating this point of view.[3] The Copenhagen interpretation implies a lack of complete determinism in that future events do not follow inevitably from past conditions; pure chance thus plays some role. Today many physicists accept this interpretation, and I personally find it completely satisfying. Some physicists, however, have strong objections to it. Einstein and Schrödinger in particular were strongly critical of the interpretation.

Einstein's objection is summarized in his often-quoted statement that "God does not play dice." In a number of forceful but always friendly arguments with Bohr, Einstein tried to devise ways of circumventing the uncertainty principle, but Bohr was always able to show that he was in error—sometimes by invoking Einstein's theory of relativity!

Schrödinger was an intensely emotional person, which often led to rather bizarre personal behavior. At the beginning of World War II he became director of the Institute for Advanced Studies in Dublin, but in 1945 he resigned after becoming involved in a quarrel about the way his office was cleaned. He usually dressed unconventionally and, as a result, had difficulty gaining admission to scientific meetings on several occasions—sometimes to lectures he was to give himself![4]

When the Copenhagen interpretation was put forward he did not just object to it but found it deeply distressing. With regard to Born's probability explanation, Schrödinger said, "I can't believe that an electron hops about like a flea." In one exchange with Bohr he said that since people were giving these interpretations to his wave mechanics, "I regret having been involved in this thing." To Schrödinger an electron has wave properties and should not be regarded as a particle darting about. In his view electronic orbitals are to be considered in terms of wave properties, not as the movement of particles.

▲▲▲

In the previous chapter we saw that the second law of thermodynamics has to be interpreted on the basis of pure chance. We have now seen again that in the new quantum mechanics, chance is all-important. We cannot say exactly where an electron is and what it is doing, except in terms of the probability of its being in a particular place. Chance is an essential ingredient of the structure of matter and of how processes occur. It is in fact concerned with everything, including ourselves and our behavior.

Although the quantization of energy did not easily reveal itself to scientists, we have seen in the present chapter that its existence helps us greatly to understand how our universe works. In the next chapter we meet another aspect of energy that was discovered shortly after the quantum theory, namely that mass and energy can be interconverted. This is one of the consequences of Albert Einstein's theory of relativity, which gives us further insight into the laws of nature.

5

# THE THEORY OF RELATIVITY

Albert Einstein, in addition to proposing in 1905 that radiation is quantized, formulated his special theory of relativity in the same year. This theory, and the general theory of relativity, which he developed in 1916, are regarded as among the greatest of all contributions to science. They have had a profound influence on the way scientists think about fundamental problems. In addition, some aspects of the theories have had remarkable practical consequences, leading in particular to an important way of producing energy by means of nuclear processes.

Albert Einstein (1879–1955; see fig. 45) was born in Ulm, Germany. His father was a rather unsuccessful electrical engineer who had constant business troubles that caused the family to move frequently. Albert entered the Swiss Federal Institute of Technology in Zurich at the age of seventeen. Partly because some of his teachers did not appreciate his attitude to them, he was unable at first to obtain an academic appointment. As a result, he had to settle for a junior and rather dull position in the patent office in Bern, Switzerland. In 1901, disliking the political atmosphere in Germany, he renounced his German citizenship in favor of Swiss citizenship. In 1903 he married Mileva Maric, who had been a fellow student of physics at the institute. The previous year, their daughter Lieserl had been born; she was probably put up for adoption and her fate is unknown. The Einsteins later had two sons.

**Figure 45.** Albert Einstein (1879–1955) was German-born but renounced his German citizenship, becoming a Swiss citizen in 1901 and an American citizen in 1940. He revolutionized physics, particularly with his theory of relativity and his theory that radiation can behave as a stream of particles (quanta, or photons). He won the 1921 Nobel Prize for physics, specifically for his work on the quantization of radiation. (Edgar Fahs Smith Collection, University of Pennsylvania Library)

Einstein did competent work in the patent office, and since the position was by no means demanding he had the leisure to carry out a good deal of research and to publish several scientific papers. The year 1905 was an *annus mirabilis* for him, for in that year he published six papers, three of which were worthy of Nobel Prizes. The first of these was on the quantization of energy, which we discussed in the last chapter and for which he received a Nobel Prize much later. Another was on Brownian movement, which is the rapid and incessant motion observed for example with smoke particles in a gas or tiny particles in a liquid. The third was on the special theory of relativity. All three papers appeared in the *Annalen der Physik*, a prominent German scientific journal, in late 1905.

▲▲▲

The special theory of relativity is based on formulating the laws of physics in a way that is common to all observers under any conditions. To understand the special theory we must know something of its background. When the wave theory of light was being developed, most investigators assumed that there was some medium that carried the light waves. Ripples on the surface of a pond are transmitted by the water, and thus it seemed necessary to postulate some medium that would transmit the light waves. This was called the *ether* (sometimes, especially formerly, spelled *aether*), and it was often called the luminiferous (i.e., light-bearing) ether. The ether was presumed to pervade all space. Although Maxwell mentioned the ether only once in his famous book on his electromagnetic theory, his *Treatise on Electricity and Magnetism* (1873), in his article "Ether" in the ninth edition of the *Encyclopaedia Britannica* (1875) he expresses very clearly his belief in the existence of an ether. However, he considered that his theory of electromagnetic radiation was valid whether or not the ether exists, or whatever its nature is. Today, now that relativity theory has been developed, the assumption that an ether exists seems unnecessary.

Various astronomical observations about the motion of the stars and planets led scientists to the conclusion that these bodies must move through the ether without producing any disturbance. If this is the case it should be possible to determine the speed with which Earth moves through the ether by making observations on the speed of light. Suppose that Earth is traveling through the ether in a particular direction with a speed of $v$ relative to the ether. If light, moving with speed $c$, travels in the same direction, its speed relative to the ether should, according to classical theory, be $v + c$. If the light travels in the opposite direction, its speed relative to the ether should be $c - v$. The situation is analogous to a woman swimming with a speed $c$ in a river flowing at the rate $v$. If she travels in the same direction as the stream, the speed relative to the bank is $c$ + v; if she swims in the opposite direction the speed is $c - v$. The relationships are easily worked out if the swimming is perpendicular to the flow, or if it is diagonal. In the case of measurements made of the speed of light in directions at right angles to each other, the theory leads to the result that the ratio of the measured rates should be the square root of $(1 - v^2/c^2)$, which we write as $\sqrt{(1 - v^2/c^2)}$.

▲▲▲

It follows that information about the ether could be obtained from measurements of the speed of light in two directions at right angles to one another. An experiment could be carried out in which a beam of light is sent out in such a way that it strikes a mirror some distance away and is reflected back—much the same procedure used by Fizeau and Foucault and described earlier. Light from the same source could also be made to travel in a direction at right angles to the previous direction, and to be reflected by another mirror. By comparing the times of arrival of the two reflected beams, it would then be possible to calculate $v$, the speed with which Earth is traveling through the ether. An experiment of this kind had been suggested by Maxwell when developing his electromagnetic theory.

The first such experiments were carried out in 1891 by distinguished American physicist Albert Abraham Michelson (1852–1931). Born in Strelno, which is now in Poland, Michelson emigrated with his parents to the United States at the age of four. He was educated at the Naval Academy in Annapolis, Maryland, and following a tour of duty at sea was appointed a naval instructor. Throughout his career he devoted much effort to improving the accuracy of the measurement of the speed of light and designed instruments for this purpose. In 1880 he took a study leave and worked in the laboratories of eminent physicist Hermann von Helmholtz at the University of Berlin. In Helmholtz's laboratory he built an instrument that allowed two narrow beams of light to be emitted in directions at right angles to one another and for the speeds to be compared in the two directions. He found that the speeds were exactly the same.

Michelson left the navy in 1881 and became professor of physics at the Western Reserve University in Cleveland, Ohio. A professor of chemistry there, Edward Williams Morley (1838–1923), shared Michelson's great enthusiasm for making accurate physical measurements. The two men therefore joined forces and together made many further measurements on the "ether-drift" problem, always finding a null result. They tried with various orientations of the beams but always found that the speed of light was independent of the direction of propagation. Their joint paper published in 1887 created a deep impression on physicists, and in the following year Michelson was awarded a national prize, the citation for it saying that it was "not only for what he has established, but for what he has unsettled."

This conclusion that the speed of light was independent of direction is now generally accepted and, as we shall see, is consistent with Einstein's theory of relativity. The result, however, did not go unchallenged. Morley later carried out some further experiments with American physicist Dayton Clarence Miller (1866–1941), who was professor at the Case School of Applied Sciences in Cleveland. Some of their results seemed to yield a positive result, suggesting the existence of an ether, but Morley discounted them as due to experimental error. Miller, on the other hand, took them seriously and from 1902 to 1932 carried out many further experiments on his own, some of them at Mount Wilson Observatory in California. His announcement in 1925 of positive results attracted much attention, since they were regarded by some as a refutation of Einstein's theory of relativity. On the basis of this work, Miller was awarded a prize by the American Association for the Advancement of Science. A reappraisal of his work in the 1950s suggested that his results were erroneous because of slight variations in the temperature of his equipment.

The negative result obtained by Michelson and Morley was generally accepted, and various suggestions were made to explain it. Similar ideas were put forward by G. F. Fitzgerald and H. A. Lorentz. Irish physicist George Francis Fitzgerald (1851–1901) was educated at Trinity College, Dublin, remaining there as professor for the rest of his short life. To explain the null result from the Michelson-Morley experiment he suggested that a body moving through the ether contracts slightly in its direction of motion, in proportion to its velocity. Such contraction would not be experimentally observable, since any measuring instrument would contract correspondingly, and the length would appear to be unchanged. The idea was that the apparatus used in the Michelson-Morley experiment would change its dimensions in such a way as to compensate exactly for the expected change in the observed velocity of light.

A similar explanation was put forward by distinguished Dutch physicist Hendrik Antoon Lorentz (1853–1928; see fig. 46), who for many years was professor of theoretical physics at the University of Leiden. He, too, suggested that the null results obtained by Michelson and Morley were due to the contraction of matter as it

**Figure 46.** Hendrik Antoon Lorentz (1853–1928) was a famous Dutch theoretical physicist whose interpretation of the Michelson-Morley experiment provided a firm basis for Einstein's more comprehensive theory of relativity. (Edgar Fahs Smith Collection, University of Pennsylvania Library)

moves relative to the ether, and this contraction is now commonly known as the *Lorentz-Fitzgerald contraction*. In 1904 Lorentz developed a mathematical treatment of the contraction, concluding that if, for example, an electron has radius $r_o$ when at rest, its radius $r$ when it is moving at speed $v$ is given by

$$r = r_o \sqrt{(1 - v^2/c^2)}.$$

Since $v$ is bound to be less than $c$, the ratio $v^2/c^2$ is a fraction, and therefore $\sqrt{(1 - v^2/c^2)}$ must be less than 1; $r$ is therefore less than $r_o$, corresponding to a contraction.

Lorentz then took this argument a stage further, deducing from electromagnetic theory that the mass of a particle such as an electron would be inversely proportional to its radius. If, therefore, we represent the mass of the electron when it is stationary as $m_o$, its mass when moving with a speed $v$ is given by

$$m = \frac{m_o}{\sqrt{(1 - v^2/c^2)}}.$$

We shall see that the same relationship was later given by Einstein in his special theory of relativity, so Lorentz's work was especially important in paving the way for relativity theory.

Einstein's great contribution in his special theory of relativity was to arrive at these and other relationships in a much more satisfactory and comprehensive way. Einstein discarded the ether concept as unnecessary, and made two significant postulates. One was that the laws of motion are exactly the same when they are determined by observers moving at different speeds. The second is that the velocity of light has a constant value, irrespective of how the source of the light is moving relative to the observer.

The ideas of the special theory can be understood as follows. Suppose that a man in a boat is drifting along a river that is flowing at speed $v$ and that a signal light on the bank ahead of the boat flashes a beam of light. To an observer on the bank the speed of the light would be $c$. The man on the boat, moving toward the signal, however, would according to classical physics measure the speed of the light relative to him as $c + v$. If he were traveling away from the light, the speed of the light would appear to be $c - v$. According to Einstein's theory, however, an observer on the boat will always find the speed of light to have the value $c$, irrespective of the speed of motion $v$. This can be visualized in terms of the Lorentz-Fitzgerald contraction; the instruments used for measuring the relative speed of light undergo a contraction. As a result of the change the observer always finds the speed of the light to be the same, irrespective of the relative motion.

Einstein's theory led to the same formula that Lorentz gave to express the relationship between the mass $m$ of a body moving with relative speed $v$ and the rest mass $m_o$ (see equation above). The change in the effective mass of an object is small if $v$ is less than about a tenth of the speed of light. If $v$ is one-tenth of the velocity of light, i.e., $v/c$ is 0.1, which means that $\sqrt{(1 - v^2/c^2)}$ is 0.995, the mass is $m_o/0.995 = 1.005\ m_o$. In other words, the effective mass is only half a percent greater than the rest mass $m_o$. If, on the other hand, the speed $v$ is 99 percent of the speed of light, the effective mass is 7.1 times the rest mass. If the speed is 99.9 percent of the speed of light, the mass is 22.4 times the rest mass. Lorentz had deduced these relativity corrections only for an electron; Einstein's derivations were more general, applying not only to an electron but to anything.

Even before Einstein presented his special theory there had been some experimental evidence for the increase in the mass of an object when it moves at high speed. In 1900 a German physicist named Kaufmann described a method for determining the ratio of the charge $e$ to the mass $m$ of an electron emitted as a beta particle from a radioactive source. These electrons are moving at high speeds, and Kaufmann obtained good evidence for a decrease in $e/m$ as the speed of the electrons approached the speed of light. When the speed of the electron was less than about one tenth of the speed of light, $e/m$ was found to be constant, but at higher speeds the decrease could be detected. Now that it has become possible to accelerate electrons to speeds that are not far short of the speed of light, this increase in mass at high speeds has often been confirmed experimentally.

From this relationship between the effective mass and the rest mass, Einstein was able to derive an expression for the relationship between mass and energy. Scientists denote the change in a quantity by preceding the symbol by the Greek capital delta, $\Delta$; a change in the mass is thus denoted by $\Delta m$. Einstein found that this change in mass for a body moving with the speed of light $c$ would be the kinetic energy divided by the square of the speed of light: $\Delta m = E_k/c^2$. Then, by a simple extension of this relationship, he proceeded, as he put it, to prove that "the mass of a body is a measurement of its energy content." In other words, when the energy of a body is changed by an amount $E$, no matter what form the energy takes, the mass of the body will change by $E/c^2$. It is thus possible to write the equation $E = mc^2$. This means that under certain circumstances, energy and mass can be interconverted. For example, a process may occur in which there is a loss of mass, in which case the gain of energy is given by this equation. Alternatively, when we accelerate a body such as an electron, we are providing it with energy, some of it going into its increased mass.

For ordinary chemical reactions this mass-energy relationship can be ignored, the mass changes being much too small to measure. For example, if 2 grams of hydrogen reacts with 16 grams of oxygen to form water (the chemical equation is $H_2 + \frac{1}{2}O_2 \rightarrow H_2O$), the heat (or energy) evolved is found experimentally to be 241,750 joules. From the equation $E = mc^2$ it can be calculated that there must be a decrease in mass of $2.7 \times 10^{-9}$ grams, a very tiny amount. Even if it were a thousand times as great as this, it would be very difficult to detect.

When a nuclear reaction occurs, on the other hand, there can be an enormous production of energy. Since nuclear reactions occur in the Sun and other the stars, they help us to understand the ultimate sources of energy. In particular, our deductions about how the universe must have been created require us to take into account mass-energy transformations as expressed by the formula.

Surprisingly, the idea of the conversion of mass into energy had been considered even earlier in the twentieth century, before the publication of Einstein's 1905 paper on the special theory of relativity. Johann Elster and Hans Geitel (the "Heavenly Twins," whom we met in the last chapter in connection with the photoelectric effect) considered the source of the energy in the radiation that was emitted by a radioactive substance. After carrying out experiments to test other possibilities, they concluded in 1903 that the energy could not come from any outside source, but must come from the atoms themselves. Because of this conclusion, the house in Wolfenbüttel where Elster, his wife, and Geitel lived and did their work bears a memorial plaque on which the two men are honored as the *Entdecker der Atomenergie,* the discoverers of atomic energy.

To get some idea of the enormous amounts of energy released when mass is converted into energy, suppose that you were skiing down a hill and attained a high speed. Suppose, hypothetically, that you reached the speed of light.[1] It follows from Einstein's formula that if you hit a tree, the liberation of energy due to your complete annihilation would be equivalent to that produced by about forty nuclear warheads. There is another way of appreciating the enormous amounts of energy that are tied up in nuclei. If all the energy in one ounce of matter could be released, it would be enough to drive one of the largest ocean liners across the Atlantic.

▲▲▲

Einstein's three papers of 1905 created quite a stir in scientific circles, but it was not for a few years that he was rewarded with an academic appointment. For four more years he continued with his work in the patent office, but in 1909 a special professorship was created for him at the University of Zurich. In 1911 he moved to the University of Prague, but in the following year he returned to Zurich, this time not to the University but to the famous Eidgenössiche Technische Hochschule, or ETH. In 1914 he was offered a professorship at the University of Berlin, a position created especially for him and conferring special privileges. At the same time he was to be director of the new Kaiser Wilhelm Physical Institute in Berlin. In addition he was offered membership of the prestigious Prussian Academy of Sciences with a salary, which was not normally offered to the members. Since Einstein had renounced his German citizenship he hesitated before he agreed to accept these appointments. Years later he was disconcerted to learn that, particularly because of his acceptance of membership of the Academy, the German government again claimed him as a citizen. He retained the appointments in Berlin until 1933, at which point the actions of the Nazis induced him to move to the United States.

His most productive years were before 1916, when his general theory of relativity was published. It is difficult to get a deep understanding of this theory without wading into the mathematics, but one can get a feeling for the theory as follows. The theory deals with three-dimensional space as well as with time, which it treats as a fourth dimension. Suppose that we ignore two of the four dimensions and visualize space-time as a two-dimensional rubber sheet. A heavy ball placed on the sheet creates a dent, and we can imagine the movement of a smaller ball on the sheet. If it comes close to the larger ball, its path curves toward it, a motion that represents gravitational attraction. In other words, the theory does not focus attention on the force itself, but regards the force as arising from the curvature of space. If space is curved, light will also have a curved path when it passes near to a massive object. We shall later discuss the curvature of the path of light from stars when it passes close to the Sun, first observed experimentally in 1919.

Einstein developed mathematically a general theory of space and time, describing the past, present, and future state of the universe. At that time it was assumed that the universe is static, the stars and galaxies being at a fixed and constant distance from us. Einstein was therefore disappointed to find that his equations were giving rise to a universe that was expanding with time. To take care of this apparent discrepancy he arbitrarily introduced into his equations an extra term, referred to as the *cosmological constant*, which allowed for the universe to be viewed as stable. Later, as will be discussed in chapter 9, it was found experimentally that the universe is indeed expanding, so Einstein's cosmological constant could be thrown away. Afterward Einstein described his introduction of the cosmological constant as the biggest blunder of his career, but here he was being too hard on himself. After all, the aim of science is to explain the workings of the universe around us, and that was Einstein's intention in introducing his constant. Presumably he felt badly about not having had the courage of his convictions by insisting that the universe must be expanding, whatever the evidence might be. As it is, his equations without the cosmological constant do predict just that, which is a great achievement. He need not have blamed himself for trying to accommodate the observational evidence as it appeared to be at the time.

The matter of the expanding universe needs a little more discussion. What the latest astronomical evidence shows is that all the distant stars and galaxies are moving away from us. This might seem to suggest that we on Earth are at the center of the universe. The picture that we get from the general theory is, however, different. Let us again visualize space-time as a stretched rubber sheet, such as the surface of a balloon that is being inflated. Suppose that representations of the galaxies have been painted as spots on the surface. As the balloon gets bigger, all of the spots of paint get farther from one another. Suppose that we choose two spots that are a certain distance from one another at a particular time, and find that after a certain time they are twice as far apart. Two other spots that were initially twice as far apart will have become four times as far apart after the inflation. An observer at any position will therefore notice that if a spot that is a certain distance away is receding at a certain speed, one that is twice as far away is receding twice as fast. American astronomer Edwin Hubble discovered, at about the same time, that the galaxies are doing just that; those that are twice as far away from us are receding twice as fast. What he observed is therefore fully explained by the general theory of relativity.

Einstein's theory of relativity has important consequences for time as well as for space. We have seen that measurements of the speed of light show that it is independent of the motion of the person measuring it. We have also seen that a body contracts when it moves, which means that space contracts as a result of the motion of the person making the measurement. Einstein realized that this leads to a remarkable consequence. Since speed is distance divided by time, it can remain the same only if the distance and the time vary with speed in exactly the same ratio. A person moving at a very high speed would find that both space and time would change in the same way, such that their ratio would remain the same.

It has now proved possible to verify this prediction experimentally. The speeds at which we are able to travel are only a tiny fraction of the speed of light, so the effects are exceedingly small and hard to detect. Jetliners travel at only about a millionth of the speed of light, and according to relativity theory they will only gain or lose about one millionth of a second per day. Atomic clocks, based on the oscillation of a cesium atom, are sufficiently sensitive to allow such small times to be measured. When tests with atomic clocks were made, those that traveled east lost a measurable fraction of a millionth of a second compared to clocks on the ground, while those that traveled west gained roughly the same amount.[2] These results thus support Einstein's prediction that time is affected by the motion of the instrument measuring it.

The 1921 Nobel Prize for physics was awarded to Einstein, and there are several curious circumstances relating to it. One relates to his wife Mileva, with whom he was not getting along well but who was reluctant to leave him. When it became fairly certain that Einstein would win the prize he made a bargain with his wife, which they kept: he would give her the prize money as compensation for her leaving him. Another odd feature was that the prize was awarded to him not for the theory of relativity but for his work on the quantum theory. In 1910 he had been nominated for the prize for his work on the special theory, but nothing ever came of this. The reason is that even after the general theory had been confirmed by the astronomical observations of 1919, there still remained some hostility toward the theory. Indeed, the Nobel committee had so many doubts about Einstein's work that the decision to give him the 1921 award was not made until 1922. Now that we have come to appreciate the theory of relativity it seems a surprising anomaly that no Nobel Prize was ever awarded for it.

▲▲▲

British astrophysicist Arthur Eddington (1882–1944) made a particular study of relativity theory, and he and others made an important contribution by providing, in 1919, the first experimental proof of the theory. We have seen that Einstein's general theory led to the conclusion that a large enough mass could exert gravitational attraction on a beam of light and cause the beam to bend. It made a clear prediction, capable of being experimentally tested, that stars appearing to be near the sun during a solar eclipse (but are actually, of course, much farther away) would be displaced from the positions that had been observed for them at other times. Until 1919 it had not been possible to confirm this prediction, but on May 29 of that year there was an unusually auspicious eclipse, one that might not have occurred again for hundreds of years. On May 29 of every year the Sun moves in front of an unusually dense group of bright stars, known as the Hyades cluster. With no solar eclipse it is impossible to make observations of these stars, because the brightness produced by the Sun makes them invisible. In that particular year there was an eclipse of the Sun on that date, and preparations for making the observations had begun two years earlier.

The Astronomer Royal of the time, Sir Frank Dyson (1869–1939), was strongly in favor of organizing expeditions to observe this eclipse. He wanted Eddington to play a leading role, since in 1912 he had led a successful expedition to Brazil to observe a solar eclipse. There was, however, a difficulty: by 1917, when plans were being initiated, conscription had been introduced in Britain, and all men with the right qualifications were eligible for the draft. Eddington was in this category: he was thirty-four, unmarried, and physically fit. He was, however, a devout Quaker and an uncompromising conscientious objector to military service. This presented a problem not only to Eddington but also to the scientific community. Brilliant young physicist Henry Moseley had volunteered for military service and in 1915, at the age of only twenty-eight, had been killed at Gallipoli. It was becoming clear that scientific work was essential to Britain's success in the war; scientists were in short supply and the scientific community was pressing for exemption for qualified scientists in order for them to be available for scientific service. Eddington was already recognized as a scientist of the first rank, and a group of distinguished scientists pressed the Home Office to give him an exemption. If this had not been granted to him he might have been sentenced to jail and at the very least would have been obliged to forsake scientific work and undertake work in agriculture or in industry.

Eventually the Home Office agreed and sent Eddington a letter to sign and return. Eddington, however, was a scrupulous man, and added a note to the effect that if he were not deferred for scientific reasons he would apply for exemption on conscientious grounds. This produced a bureaucratic problem, and some of the scientists who had supported Eddington were annoyed with him for creating what they thought to be an unnecessary difficulty. The result was a further round of discussions, with the upshot that Eddington was deferred provided that he would lead the expedition—which of course is what he wanted to do!

Two sites were chosen for the astronomical expedition: one was the island of Príncipe, off the African coast, just north of the Congo. Eddington went there to lead a Cambridge team, and to reduce the odds of failure due to bad weather, a second British expedition, headed by astronomer C. R. Davidson, went to a site in Sobral in northern Brazil and used a somewhat different photographic technique. It was just as well that they went there, because it rained at the African site, and only two satisfactory plates were obtained. Eddington arranged for these to be developed and examined on the spot, before they were transported back home. Unfortunately, during their transportation from Africa back to Cambridge the plates suffered from heat and humidity and were difficult to analyze further. As a result, it was not until November 1919 that it could be announced to the world that the observed bending of the light agreed with Einstein's predictions to well within the margin of experimental error.[3] Now that astronomers can make observations of certain stars using radio waves it is much easier to make similar measurements, since it is no longer necessary to wait for a total eclipse. Many investigations have now confirmed this prediction of Einstein's theory.

No scientific experiment has ever generated so much excitement, not only in the scientific world but among the public. The official announcement was made on November 6, 1919, at a joint meeting of the Royal Astronomical Society and the Royal Society at Burlington House, Piccadilly. Dyson spoke first, followed by Eddington, and the President of the Royal Society then summed up by saying that "This is the most important result obtained in connection with the theory of gravitation since Newton's day."

The excitement rapidly spread to the general public. After the strain of World War I the time was particularly ripe for an event that transcended blind nationalism. Here was an important scientific theory proposed by a German-born Jewish scientist being confirmed by Englishmen so soon after a great war between their two countries. Einstein instantly became the best-known media celebrity on the planet. A headline in the London *Times* of November 7, 1919, announced, "Revolution in Science: New Theory of the Universe: Newtonian Ideas Overthrown." There followed a detailed account of the Burlington House proceedings. Headlines in the *New York Times* for November 10, 1919, screamed, "Light All Askew in the Heavens: Men of Science More or Less Agog over Results of Eclipse Observations," and, "Einstein Theory Triumphs: Stars Not Where They Seemed or Were Calculated to Be, but Nobody Need Worry."

The *New York Times* detailed one of its reporters, Henry Crouch, to cover the news of the event in England, but he was their golf reporter and had little understanding of science. Also, in the interests of sensational journalism he was not at all inhibited about taking liberties with the truth. One of the headlines for his stories ran, "A Book for 12 Wise Men: No More in All the World Could Comprehend it, Said Einstein When His Daring Publishers Accepted It." This headline had less relationship with reality than most, which is saying a lot. Einstein was not writing a book at the time and was dealing with no daring or even cowardly publishers. He would never have said that only twelve people could comprehend the theory of relativity, since he knew that by this time many physicists and astronomers understood it well.

It may be that it was this headline that instilled the idea among the public that Einstein's theory was incomprehensible even to most scientists. A story about Eddington dates from about that time. Someone said to him that he had been told that including Einstein himself there were only three people in the world who understood his theory. Eddington did not immediately reply and was then asked if he agreed with the statement. "Well," said Eddington, "I've been trying to think who the third person could possibly be." It is most unlikely that Eddington ever said any such thing, but the story is such a nice one that I continue to tell it whether it is true or not.

Einstein himself was not impressed by all the fame that had come to him and resisted it as much as he could. Two weeks after the public announcement of the confirmation of his theory he wrote a letter to the London *Times* in which he commented rather sardonically that now that his theory had been vindicated the Germans were proudly calling him a German man of science and the English were calling him a Swiss Jew. If his predictions had been wrong so that he had been shown to be a *bête noire*, he said, the Germans would have called him a Swiss Jew and the English would be calling him a German man of science.

In 1934 he accepted an appointment at the Institute for Advanced Study at Princeton, where he remained to the end of his life, continuing to struggle with the meaning of the physical world. His work attracted much interest among the public, and for many years he was probably the best-known scientist in the world. When I was a graduate student at Princeton just before the war, I often passed him walking along the edge of the golf course on his way to the Institute. He always grinned and gave an amiable wave. His appearance was usually of the unconventional kind always associated with him, its particular feature being no tie (considered highly eccentric at the time!) and no socks.

Many stories were told about him by the Princeton staff and students, and I do not know how many of them were true.[4] One relates to the first arrival in Princeton of Einstein and his second wife Elsa. A professor who was a neighbor called on him one day when he was out and was greeted by Mrs. Einstein. The professor said that he was a chess player and that he had heard that Professor Einstein also enjoyed playing chess; he would therefore be very pleased if they could play together some time. "Certainly not," replied Mrs. Einstein. "The professor would never want to do anything like that," and she firmly shut the door. The visitor was nonplussed until he finally learned the reason for the misunderstanding: Mrs. Einstein had thought that Professor Einstein was being invited to play *jazz*, which she herself pronounced "chess." Einstein played only classical music on his violin.

Another incident occurred during the time I was in Princeton. The secretary to the Institute received a telephone call from someone with a German accent who asked for Professor Einstein's address. The secretary said that she was sorry, but the professor had given strict instructions that his address was not to be given out. "Now that is very difficult," said the caller. "You see, *I* am Professor Einstein and I have forgotten where I live and I want to go home."[5]

Einstein often had difficulty going home. For some time after the Einsteins arrived in Princeton, neighbors would sometimes go into their own sitting rooms and find Einstein calmly installed reading a newspaper; he had "come home" to the wrong house. Consideration was given by the neighbors to painting their front doors a different color from that of the Einsteins so that the professor could more easily find his own house.

Einstein was always rather vague about trivial details like paying bills or filing income tax forms; fortunately he always had a secretary to deal with such matters. The US income tax form ends with the following questions: "1. Did anyone help you with completing this return? 2. What was the nature of the help?" To 1, Einstein answered, "My secretary"; to 2, "She told me what to write."

▲▲▲

When the equation $E = mc^2$ was first proposed, there were inadequate data for testing it. Today there are many examples, in the form of radioactive disintegrations and other nuclear reactions. We will look at just a few examples.

Ordinary radium undergoes the following process, the products being the elements helium (He) and the gas radon (Rn):

$$^{226}_{88}\mathrm{Ra} \rightarrow {}^{4}_{2}\mathrm{He} + {}^{222}_{86}\mathrm{Rn}.$$

The following are the masses of the nuclei involved:

$$^{226}\mathrm{Ra} \quad 226.0312 \text{ u};$$
$$^{4}\mathrm{He}\ 4.0026 \text{ u} + {}^{222}\mathrm{Rn} \quad 222.0233 \text{ u} = 226.0259 \text{ u}.$$

The unit *u* is the *atomic mass unit*; it is defined in such a way that the mass of the isotope $^{12}C$ (carbon-12) of the carbon atom is exactly 12 u. We see that the mass has decreased by 226.0312 – 226.0259 = 0.0053 u. It is easy to calculate from the formula $E = mc^2$ that one atomic mass unit corresponds to an energy of 931.5 million electron volts (MeV), and it follows that a decrease in mass of 0.0053 u will lead to the production of 4.9 MeV of energy. The *electron volt* (eV) is a convenient unit to use in work of this kind; it is the energy acquired by an electron when it passes through a voltage drop of one volt. One electron volt is roughly the energy that is carried by one photon of visible light. We get some idea of the vast amount of energy evolved in these nuclear transformations when we note that in ordinary chemical reactions, even in conventional explosions, the energy produced is only a few electron volts. With nuclear processes we are talking about millions of electron volts.

Two other types of nuclear transformation are of great importance. The first is called *nuclear fission* and involves the breakdown of a nucleus into smaller nuclei. The first nuclear fission reaction was brought about in 1932 in the Cavendish laboratory at Cambridge by two of Rutherford's assistants, John Douglas Cockcroft (1897–1967) and Ernest Thomas Sinton Walton (1903–95). In the Cockcroft-Walton experiment a beam of protons of high energy was directed at a target of the element lithium. They found that alpha particles (helium nuclei) were released, and the process that occurred was found to be

$$^{7}_{3}\text{Li} + ^{1}_{1}\text{H} \rightarrow ^{4}_{2}\text{He} + ^{4}_{2}\text{He} + 17.2\ \text{MeV}.$$

(The reader might like to check this equation. The nucleus $^{7}_{3}\text{Li}$ contains 3 protons and 4 neutrons, giving it an atomic mass of 3 + 4 = 7. When this reacts with a proton, $^{1}_{1}\text{H}$, an intermediate species containing 4 protons and 4 neutrons is formed. This, however, is highly unstable and at once divides into two nuclei, each of which contains 2 protons and 2 neutrons and is designated $^{4}_{2}\text{He}$. At the same time 17.2 MeV of energy are liberated.) Rutherford, on hearing of this experiment, said that it was "the most beautiful sight in the world." When Cockcroft later suggested to Rutherford that reactions of this kind might eventually be used for the commercial generation of power, Rutherford said that the idea was "all moonshine," but within a few years he changed his mind. Cockcroft, a genial and decisive man, had a distinguished career, later becoming Jacksonian Professor at Cambridge, founding director of the Atomic Energy

Research Establishment at Harwell, and the first master of Churchill College, Cambridge. He was a leading statesman of science, combining research and administrative skills.

Another important nuclear reaction is brought about by bombarding uranium-235 with neutrons. Uranium-235 captures a neutron with the formation of uranium-236, which breaks down into two nuclei, barium-145 and krypton-88. Since 145 + 88 = 233, 3 less than 236, there are three neutrons left over. The neutron, with unit mass but no charge, is written as ${}^{1}_{0}n$, and the process is

$${}^{235}_{92}U + {}^{1}_{0}n \rightarrow {}^{236}_{92}U \rightarrow {}^{145}_{56}Ba + {}^{88}_{36}Kr + 3{}^{1}_{0}n + \text{energy.}$$

The amount of energy released in this process is large, about 200 MeV. What is particularly noteworthy about this reaction is that a single neutron has produced a large amount of energy without any loss of neutrons from the system; in fact, three neutrons have appeared in its place. These neutrons will cause the breakdown of neighboring uranium nuclei, with the production of more neutrons. When this occurs we have what is called a *chain reaction,* and within a fraction of a second many of the uranium nuclei present will have undergone fission. A small number of neutrons introduced into uranium-235 will therefore lead to the production of vast amounts of energy in a very short time. In practice, the amount of energy released is never as much as it would be if all of the nuclei underwent fission, only a few percent of the energy that is latent in the U-235 being released. Nevertheless, the energy produced is vastly more than in an ordinary chemical reaction, perhaps a million times more.

This reaction made the commercial production of nuclear energy possible, and it has an interesting history. In December 1938, less than a year before World War II began, Otto Hahn (1879–1968) and Fritz Strassmann (1902–80), working at the Kaiser Wilhelm Institute in Berlin, observed that when uranium, the heaviest element then known, was bombarded with neutrons, one of the products was barium, a relatively light element. Soon afterward Lise Meitner (1878–1968) and her nephew Otto Robert Frisch (1904–79) interpreted Hahn and Strassmann's observation as indicating that the neutrons were breaking the uranium nucleus into two nuclei of roughly equal mass. At the time both Meitner and Frisch were refugees from Germany; Frisch was working in Niels Bohr's laboratory in Copenhagen, and Meitner (who had previously worked with Hahn) was working in Stockholm. They also calculated that during the process there was a substantial loss of mass and therefore, by Einstein's relationship, an enormous liberation of energy. The energy produced was much greater than in any nuclear process that had been previously observed.

When Frisch told Bohr about their conclusion, Bohr struck his forehead and exclaimed, "Oh, what fools we have been! We ought to have seen that before." Frisch did not at first know what to call the type of process that was occurring, and asked a biologist what term was used to describe the splitting of a biological cell into two daughter cells. *Fission* was the answer, and in the Meitner-Frisch publication in the scientific journal *Nature*, the word *fission* appeared—not yet *nuclear fission*, which came later.

In Paris in the same year, Frédéric Joliot-Curie (1900–58) and his wife Irène Joliot-Curie (1897–1956)—the daughter of Marie Curie, famous for her early work on radioactivity (see chap. 1)—discovered something of great significance about the fission process. It was they who found that there were more neutrons released in the process than had been used in the initial bombardment of the uranium. These secondary neutrons could disintegrate more uranium atoms and thus set off a self-perpetuating chain reaction that would spread through the whole of the uranium present.

There is another kind of nuclear process, called *fusion*, which in principle leads to the production of even greater amounts of energy. In a fusion reaction the nuclei simply combine together. For example, the hydrogen isotope deuterium ($^{2}_{1}H$) has a mass of 2.0141 u, whereas that of a helium atom ($^{4}_{2}He$) is 4.0026 u. If therefore two deuterons are forced together to form a helium nucleus, there is a mass loss of $4.0282 - 4.0026 = 0.0256$ u and a corresponding release of energy of about 23.8 MeV:

$$^{2}_{1}H + ^{2}_{1}H \rightarrow ^{4}_{2}He + 23.7 \text{ MeV}.$$

This reaction took place extensively in the big bang, at the creation of the universe, and subsequently has occurred in stars. Its occurrence accounts for the fact that there is so much helium present in the universe. Of all of the matter present in the universe, one-quarter (by weight) is composed of ${}^{4}_{2}He$ atoms; about three-quarters is hydrogen, with only traces of all the other elements.

When attempts are made to make this reaction occur on Earth, a difficulty arises from the fact that the highest temperatures that can be attained are much lower than those in the big bang or in stars and are too low for ${}^{4}_{2}He$ to be formed readily. Instead, the following two reactions, which produce less energy, occur preferentially:

$$ {}^{2}_{1}H + {}^{2}_{1}H \rightarrow {}^{3}_{2}He + {}^{1}_{0}n + 3.28 \text{ MeV}; $$
$$ {}^{2}_{1}H + {}^{2}_{1}H \rightarrow {}^{3}_{1}H + {}^{1}_{1}H + 4.04 \text{ MeV}. $$

These are the reactions that occur for the most part in the so-called *hydrogen bomb*, or *thermonuclear bomb*. Another difficulty with the reactions is that the two positively charged nuclei repel one another and resist being brought together. Special techniques are therefore necessary, particularly to produce extremely high temperatures; it is because of these high temperatures that the processes are called *thermonuclear*. In stars the temperatures are sufficiently high that the fusion reaction to form ${}^{4}_{2}He$ is going on all the time. The enormous amounts of energy released in the fusion reactions keeps them hot for billions of years, but eventually stars die when their nuclear fuel is spent.

▲▲▲

Relativity theory differs from the other great scientific theories in that it does not enter into the lives of most of us. Newton's theories of mechanics are relevant every time we throw a ball or an apple falls from a tree. Darwin's theory of evolution is relevant to any interests we may have in the development of the human race or even in gardening. Planck's quantum theory, whether we know it or not, relates to anything we see on a television screen or any time we use a computer.

With Einstein's theory, however, the situation is different. We may receive some of our power from nuclear power plants, but aside from that we cannot point to much that we see in our daily lives and say that it is so because of relativity. Even scientists for the most part go through their entire research careers without making any use of Einstein's theory. In my own research in physics and chemistry I never had to use it, and I only know a few situations where relativity has to be introduced into a chemical problem. One of them relates to the color of gold. When calculations were first made on the structure of solid gold, the results suggested that gold should have the same color as silver, which would have been disappointing for lovers of precious metals. Obviously something was wrong with the calculations, and it was traced to the fact that in gold, because of the high charge on the atomic nucleus, the electrons move at speeds approaching that of light. A relativity correction was therefore necessary, and when it was introduced the gold color could be understood. It is rather fascinating to think that the beauty of a gold object is related to the theory of relativity. That is the only thing I know of that relates relativity to our daily lives.

Another curious thing about relativity theory is that it has received opposition from so many people who have never taken the trouble to understand it. This is illustrated by a curious incident in 1934, when Einstein spent some time in England and was invited to give a lecture on his theory at Oxford University. Present were a number of distinguished philosophers who, to Einstein's surprise, attacked the theory on philosophical grounds.[6] Einstein explained in vain that his theory was a scientific theory, to be judged solely by how it fit the experimental information; it makes no sense to criticize it on any other grounds.

The same, of course, is true of the theory of evolution, which has often been criticized on religious grounds. Religion played a role in another interesting incident on the same visit of Einstein to England. He was the guest of honor at a dinner party attended by many distinguished people, including George Bernard Shaw, the archbishop of Canterbury, and J. J. Thomson. Apparently the archbishop, Dr. Randall Davidson, had been told by someone (according to one account it was Thomson) that he should pay attention to Einstein's theory, as it might affect his religious belief. The archbishop was worried, since he had found himself unable to understand the theory. At this dinner he found himself sitting next to Einstein and discussed the matter with him. He was relieved when Einstein said rather wearily, "Do not believe a word of it; my theory is a purely abstract scientific theory and has nothing whatever to do with religion."

6

# THE SUBATOMIC UNIVERSE

Up to now we have been assuming that matter is ultimately composed solely of electrons, protons, and neutrons. We have seen that atoms have nuclei that are made up of various numbers of protons and neutrons, the chemical nature of the atoms depending on the number of protons. The nuclei are very much smaller than the atoms themselves, the space around the nuclei being occupied by orbital electrons.

This is the concept of matter that became established in the early years of the twentieth century, and for many purposes it is entirely satisfactory. Much of the research work in all of the sciences, including physics, can be carried out effectively from this point of view. Chemistry, for example, is concerned with the structure and properties of molecules and, except for some rather special applications, particles other than electrons, protons, and neutrons do not enter into it.

During the twentieth century, however, physicists began to discover other particles, some of them smaller than protons, and some even much smaller than electrons. The total number of the particles that have been identified is now well over a hundred. Some of the particles observed are of little significance. Much work is now being done with giant accelerators, in which large amounts of energy are used to smash atomic nuclei into fragments. When one smashes something to pieces, one inevitably finds many types of fragments, most of which cannot be regarded as the component parts of what was smashed up. Only a few of the subatomic particles identified can be regarded as the building blocks of matter, and our attention will be mainly limited to them.

Particles have been classified in several different ways, and some types are shown in Table 1. The hadron-lepton classification is a particularly important one; in brief, hadrons form the nuclear matter, such as protons and neutrons, whereas the leptons are electronic matter. The fundamental distinction between hadrons and leptons is

**Table 1. Types of subatomic particles**

| *Type* | *Word origin* | *Properties* | *Examples* |
|---|---|---|---|
| Hadrons | Greek *hadros*, 'bulky' | Strongly interacting | Protons, neutrons, mesons |
| Leptons | Greek *leptos*, 'small, delicate' | Weakly interacting | Electrons, muons, tauons, neutrinos |
| Mesons | Greek *mesos*, 'middle' | Associated with strong force; integral spins | Pions |
| Fermions | Fermi & Dirac | Half-integral spins | All leptons, protons, neutrons |
| Bosons | Bose & Einstein | Zero or integral spins | Photons, mesons, deuteron, $^4$He |
| Baryons | Greek *barus*, 'heavy' | Mass equal to or greater than that of a proton | All nuclei; the term is sometimes restricted to fermions |
| Quarks | From Joyce's *Finnegans Wake* | Constituents of baryons | Various "colors" and "flavors": up, down, strange, etc. |

*Note:* The classification of the particles has changed somewhat over the years, and this can cause some confusion. Originally particles were classified according to their mass, the hadrons being heavier and the leptons lighter. It is now recognized that the mass is of less significance than the type of interaction.

based on the nature of the interactions to which the particles respond. Here we must digress a little to explain these interactions. There is a close relationship between the fundamental particles of nature and the forces of nature. According to Maxwell's electromagnetic theory, for example, the charge of an electron is intimately related to the force exerted by the electromagnetic field that the electron generates as a result of its charge. According to modern ideas, a particle can be regarded as essentially a source of force, and conversely a field of force is intimately related to a type of particle. We have seen that electromagnetic radiation has particles associated with it that we call photons. Similarly, according to modern theory a gravitational field is associated with particles known as gravitons, which have proved elusive up to now. One way of looking at the particle-force relationship is to note that if a particle were acted on by no force, it would be impossible to obtain any evidence of its existence, and therefore it would have no meaning as a scientific entity. Similarly, a field of force, that acted on no particle would be undetectable and thus have no scientific significance. Investigations into the nature of particles and the nature of forces must thus go hand in hand.

▲▲▲

A force is intimately related to a changing potential energy, so that associated with every type of force there is a corresponding type of energy. The following are the four fundamental types of force, in order of increasing strength:

1. The force of *gravity* (or gravitation)
2. The *weak nuclear* force
3. The *electrical* force (more precisely, the *electromagnetic* force)
4. The *strong nuclear* force.

The gravitational force is by far the weakest of them all. If we set the strength of the gravitational force at 1, in very approximate terms the strength of the weak force is $10^{25}$; that of the electric force, $10^{37}$; and that of the strong force, $10^{39}$.

We have already met the first and third of these forces. Gravity is the force that is most familiar to us in our everyday lives, and it has been recognized for many centuries. It is responsible for maintaining the planets in their orbits around the Sun, and it prevents us from floating away from Earth. The electrical force, first recognized only about two centuries ago, is also now familiar to us. It is involved in lightning storms, it lights our cities, and it drives all the convenient devices of modern life, such as telephones and computers. Just as the mass of an object determines how much gravitational force it feels and exerts, the charge on an object determines how the object responds to electromagnetic attraction and repulsion.

The gravitational and electromagnetic forces resemble one another in obeying the inverse square law, but they differ in one respect. The gravitational force is always attractive, while the electromagnetic force is attractive only when the charges are of opposite sign (e.g., for an atomic nucleus and an electron); it is repulsive when they are of the same sign.

We can appreciate the weakness of the gravitational force compared to the electromagnetic force by noting that a nail can easily be picked up by a tiny magnet, even though the nail is being attracted by the enormous mass of the planet beneath it. An apple can be held to a tree by a thin stalk, which is strong enough to counterbalance the force of gravity. When scientists carry out quantum-mechanical calculations on atoms, taking into account the electrical forces between atomic nuclei and electrons, they can completely ignore gravitational forces. By contrast, gravity is of great importance when we are considering interactions between large bodies such as stars and planets, and then we can completely neglect the electromagnetic forces, because these bodies are electrically neutral: the electromagnetic forces produced by all their nuclei and electrons cancel out.

Both the electromagnetic and gravitational forces are called long-range forces, since they act not only between bodies that are close together but between ones that are farther apart. It is true that the force decreases with distance—in fact, it decreases with the square of the distance, so that if the distance separating two bodies is doubled, the force is four times smaller. The short-range forces, on the other hand, decrease much more strongly with the separation, such that a body acts only on its immediate neighbors. In addition to the nuclear forces there are other short-range forces, such as the attractive forces that exist between water molecules, holding them together and allowing water to exist as a liquid. One example of how short-range forces differ from long-range ones is shown by the following comparison. If we compare two planets such as Earth and the much larger Jupiter, we find that everything is much heavier on Jupiter, because that planet, on account of its greater mass, exerts a stronger attraction on anything on its surface than does Earth. The material we try to lift from Jupiter is therefore being attracted not only by the material immediately next to it but by all the material in the planet.

This is to be contrasted with the energy involved if water evaporates from a large lake, as compared with the evaporation of the same amount of water from a small lake. It turns out that the amounts of energy required are exactly the same for a large lake as for a small one, because the forces holding molecules together in a liquid are short-range forces. It is only the immediate neighbors that exert any appreciable attraction; the size of the lake makes no difference.

The other two fundamental forces of nature, the strong and the weak nuclear forces, were discovered much more recently than the gravitational and electromagnetic forces. They are much less familiar to us because, being short-range forces, their strength is negligible over all but subatomic distances. When physicists began to study the forces that bind together the protons and neutrons (each of which we call a *nucleon*), they observed that the energy required to remove a single nucleon from a nucleus depends very little on the size of the nucleus. This result indicates that there must be powerful short-range forces holding the nucleons together in the nucleus. Consider a nucleus like that of the metal bismuth, $^{209}_{83}Bi$, which consists of 83 protons and 126 neutrons. The radius of the bismuth nucleus is about $6 \times 10^{-15}$ m, which is about ten thousand times smaller than that of the atom itself. With 83 protons packed into such a small space, the electrostatic repulsions are enormous. Since the gravitational attractions between the particles are negligible compared with these repulsions, there must be some strong short-range forces of attraction between the particles, strong enough to hold them together in a stable nucleus. Calculations show that these attractive forces must be about a hundred times stronger than the electrostatic force. They are usually referred to simply as the strong force. However, they are so short-ranged that they act only within a nucleus; they have no appreciable effect even on the orbital electrons that exist in the atoms.

The first important theory of the short-range strong force was put forward in 1935 by Japanese physicist Hideki Yukawa (1907–81), who was at Kyoto University. He suggested that just as electromagnetic attraction is brought about through the agency of the electromagnetic field, the attractive nuclear force is also transmitted by a field. There is, however, an important difference between the fields as far as the associated particles are concerned. Electromagnetic attractions depend on the exchange of photons between the interacting bodies, these photons having zero mass. The particle related to the strong force, however, according to the theory, has a finite mass. Yukawa estimated on the basis of his theory that this particle has a mass of about two hundred times the mass of an electron. The particle is called a *meson,* and being concerned with the strong forces, it is classified as a hadron (refer to Table 1).

This theoretical prediction led nuclear physicists to try to discover the particle experimentally. In 1936 American particle physicist Carl David Anderson (1905–91) was studying cosmic rays at the California Institute of Technology (Caltech). These rays are highly energetic beams coming from outer space and consisting of protons, electrons, and other particles. He was using the so-called cloud-chamber technique, in which rays are detected from the tracks that are produced by causing water vapor to condense on the particles. He deflected the particles by a magnetic field and observed a particle that had the same negative charge as an electron but that was about 130 times as heavy. At first this seemed to be the meson postulated by Yukawa, but further studies showed that it did not interact with atomic nuclei in the right way; it was not associated with the strong nuclear force but with the weak one. Anderson's particle was later called a *muon,* and since it is concerned with weak and not strong interactions it belongs to the class of leptons. A little later another cosmic-ray physicist, Englishman Cecil Frank Powell (1903–69) of the University of Bristol, was studying cosmic rays using another technique, recording the tracks on a photographic plate. In this way he discovered in 1947 a new particle, with a mass 274 times that of the electron. This particle had the properties predicted by Yukawa, and it was called a pi-meson (symbol $\pi$), or a *pion*. It was later found that Anderson's muon was produced when a pion decayed.

As a result of this confirmation of his theory Yukawa was awarded the 1949 Nobel Prize for physics, and Powell was awarded the prize in 1950. Anderson had already received a Nobel Prize for physics in 1936, the year in which he discovered the muon, but that was for his earlier discovery of another particle, the positive electron, or *positron*. That was also as a result of a theoretical prediction, this time made by Paul Dirac, whom we recall from chapter 4 as one of the founders of quantum mechanics. On the basis of his theory he predicted that all of the fundamental particles have their counterparts in the form of what is called *antimatter*; particles of antimatter are called *antiparticles*. What Dirac found was that the equations he developed always had two solutions. A simple way in which an equation has two solutions is if we have to take the square root of a number; the square root of 4, for example, is not only 2, it is also –2. In the same way, Dirac found that his equations that led to the existence of a negatively charged electron also led to the possibility of a positively charged electron, or positron, having exactly the same properties except for the charge. The positron was discovered experimentally by Anderson in 1932, using the cloud chamber technique, and it was for this work that he received his 1936 Nobel Prize.

Dirac's conclusion had the effect of doubling the number of elementary particles at once, and many of them have now been discovered and studied.[1] If a particle has an electric charge its antiparticle has the opposite charge; the electron has a negative charge, and the positron, its antiparticle, a positive one. If a particle meets its antiparticle, both are annihilated, their joint masses being converted into the energy of electromagnetic radiation. The universe contains a preponderance of one of the forms of matter, and any antimatter that is produced has a very short life, since it will soon meet its antiparticle and be annihilated.

An antiparticle has the same mass, spin, and lifetime as the particle itself but differs in one of several properties, such as charge or properties called "color" and "strangeness," which will be referred to later in this chapter. In 1985 a team of researchers succeeded in making complete antiatoms of hydrogen, in which the proton was replaced by the negatively charged antiproton and the electron was replaced by the positively charged positron. These atoms, however, survived for only about 40 billionths of a second, since when one of them encountered an ordinary hydrogen nucleus they were both at once annihilated. This work was done at the Organisation Européenne pour la recherche nucléaire, usually known as CERN, whose offices straddle the Swiss-French border near Geneva.[2] This organization is famous for its excellent scientific work and also for the fact that the World Wide Web was developed there, originally for the transfer of data between particle physics research groups. It now gives millions of people access to cyberspace shopping malls and to a plethora of information of all kinds.

The weak nuclear force is concerned in subtle ways with radioactive decay. Its strength is about one ten-million-millionth ($10^{-13}$) of that of the strong force, but it is nevertheless much stronger than gravity—about $10^{25}$ times as strong. The weak interaction is what is involved in the conversion of a neutron into a proton, with the emission of an electron. According to modern theory this involves the creation of a particle called the $W^-$ particle, which belongs not only to the class of leptons but also to the class of bosons (refer to Table 1). From this point of view the decay of a neutron, which is what happens in beta decay, can be represented as follows:

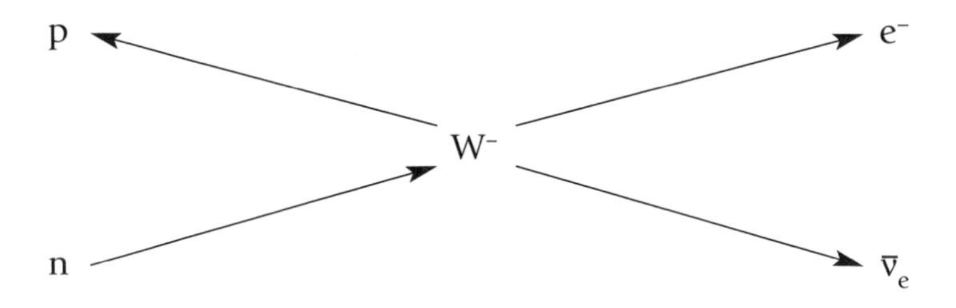

In the first stage the neutron forms a $W^-$ particle and becomes a proton. The $W^-$ particle has a lifetime of only about $10^{-23}$ seconds and at once is converted into an electron. At the same time it forms what is called an antineutrino, $\bar{\nu}_e$.

The $W^-$ particle that was predicted theoretically to occur as an intermediate in this process was first detected experimentally in 1985 at CERN. There are three bosons concerned with the weak interactions. Besides $W^-$, there is $W^+$, both having one unit of charge, and there is also a so-called Z particle, which is neutral. Each of the two charged bosons, $W^-$ and $W^+$, has a mass of about 82 MeV/$c^2$; the neutral boson is a little heavier, having a mass of about 90 MeV/$c^2$.[3] Because they are intermediates in nuclear processes they are sometimes called *intermediate bosons*.

We should now return to considering the various classification systems for particles (refer to Table 1). The basic classification is in terms of their behavior toward interactions. Particles that experience the strong interactions, which are concerned with holding them together in nuclei, are called *hadrons*; those not influenced by the strong nuclear force are called *leptons*. Some of the more important hadrons are listed in Table 2. By and large hadrons are heavier than leptons, but mass is not the basis of this classification. Certain hadrons that are lighter than most of the others are called *mesons*, the name being originally intended to signify a particle that has a "medium" mass, between that of the electron and the proton.

Table 3 lists some of the more important particles that are involved with the weak interactions, which are therefore called leptons. One of them is the particle associated with electromagnetic radiation and is called the photon; it has the special property of having no relativistic rest mass, by which we mean that its mass derives entirely from its motion; it always travels at the speed of light. The electron and its antiparticle the positron are also leptons. There is also an important group of leptons called neutrinos, but it is convenient to deal with them in a separate section later in this chapter.

Another useful classification of particles is in terms of their spin, and some explanation of particle spin is needed first. In order to explain the behavior of electrons in certain kinds of experiments it

**Table 2. Hadrons** (particles affected by the strong nuclear force)

| *Particle* | *Word origin* | *Discovered or suggested* | *Charge* | *Mass, in MeV/c²* *(no. electrons)* |
|---|---|---|---|---|
| Proton, p | Greek *protos*, 'first' | Rutherford, 1920* | +1 | 938.3 (1,837) |
| Neutron, n | Latin *neuter*, 'neither' | J. Chadwick, 1932 | 0 | 939.6 (1,839) |
| Pion, π | Greek letter *pi* | C. Powell, 1947 | +1, −1, 0 | 140 (274) |
| Quark | From Joyce's, *Finnegans Wake* | Gell-Mann, 1961 | +⅔, −⅓ | |

*Rutherford first named the proton in 1920, but its existence had previously been discussed by others, notably J. J. Thomson in his experiments on positive particles in discharge tubes.

*Note:* The charge on the particle relates to +1 for the charge on the proton, which is exactly the same, with the sign changed, as the charge on the electron. For an explanation of the mass unit used, see note 3. The numbers in brackets represent the mass in terms of number of electrons.

proved necessary to conclude that they have a property that it is convenient to call *spin*. For example, electrons were found to behave in a magnetic field in a way that can be visualized as if they were spinning in two different directions, which can be called clockwise and counterclockwise. The spinning is quantized in that the electrons can spin in only two different ways, in one direction or the other. When Dirac first worked through the quantum mechanics of an electron he found that a property called the *magnetic moment* could be assigned to a spin and that in terms of a certain significant unit of magnetic moment (in fact equal to Planck's constant divided by 2π) the possible values were +½ and −½. These numbers are referred to as the spin quantum numbers and are usually employed whenever we are talking about the spin of a particle.

**Table 3. Leptons** (particles affected by the weak nuclear force)

| *Particle* | *Word origin* | *Discovered or suggested* | *Charge* | *Mass, in MeV/c²* *(no. electrons)* |
|---|---|---|---|---|
| Electron, e | Greek *elektron*, 'amber' | J. J. Thomson, 1897 | –1 | 0.51 (1) |
| Muon, μ | Greek letter *mu* | C. Anderson, 1936 | +1, –1 | 105.7 (207) |
| Tau, τ | Greek letter *tau* | | +1, –1 | 1,784 (3,588) |
| Neutrinos, ν (e-neutrino, $\nu_e$, τ-neutrino, $\nu_\tau$, μ-neutrino, $\nu_\mu$) | Italian, 'little neutral one' | F. Reines & C. L. Cowan, 1956 | 0 | $< 2 \times 10^{-6}$ ($< 2$ eV/$c^2$) |

*Note:* The numbers in brackets represent the mass in terms of number of electrons.

There are other particles, including protons and neutrons, that have spins of +½ and –½, and others that have spins of +³⁄₂ and –³⁄₂, and so forth. Spins that are fractions like ½ rather than whole numbers are called nonintegral spins, and particles that have nonintegral spins are referred to as *fermions*. Particles that have zero spin or integral spins, such as photons (which have zero spin), are called *bosons*.

Hadrons that have masses equal to or greater than that of the protons and neutrons are often called *baryons*. All atomic nuclei are therefore baryons in a broad sense, but some scientists confine the word to those particles that have half-integral spins. An interesting point about the baryons is that whatever occurs, their number seems always to be conserved. When, for example, a neutron decays, it turns into a different baryon, a proton. As we saw earlier, in so doing it creates an electron and at the same time an antineutrino, $\bar{\nu}_e$, both of which are leptons; there is therefore conservation of leptons. There seems to be a law of nature that just as mass and energy are conserved in the universe, so are the numbers of baryons and leptons.

▲▲▲

Neutrinos form a particularly interesting group of leptons that deserve a separate section. The evidence for the existence of these particles was at first indirect, or circumstantial. Experimental studies had been made of numerous radioactive processes in which there was beta decay, which means that a beta particle, namely an electron, is emitted by one nucleus with the formation of another. When measurements were first made, in the early years of the twentieth century, of the energies of the emitted neutrons, there was an unexpected result. It had been assumed that all of the electrons would be emitted with the same energy, this energy being the difference between the energy of the emitting nucleus and that of the product. Some electrons did indeed appear with a certain maximum energy corresponding to this difference, but many appeared with lower energies. How was it possible for electrons to appear with lower energies—where was the missing energy?

There were just two possibilities: either energy was not being conserved, or there was some other particle that was emitted at the same time and that carried away the missing energy, in some manner that had hitherto escaped detection. The principle of conservation of energy had been established by so many experiments over a long period of time that no one took seriously the idea that it could be violated when a radioactive process occurs. The other explanation was therefore favored, and in 1931 Austrian-born physicist Wolfgang Pauli (1900–58) suggested that the energy was being carried away by a particle that has no charge and is so light (possibly being completely without mass, like the photon) that it is difficult to detect. Italian-born physicist Enrico Fermi (1901–54) gave the name *neutrino* (Italian for "little neutral one") to this elusive particle.

Experimental evidence for the existence of neutrinos was finally obtained in 1956 by two American physicists, Frederick Reines (b. 1918) and Clyde Lorrain Cowan (1919–74). Working with collaborators from the Los Alamos Scientific Laboratory, they used sensitive detectors placed near a powerful reactor in which fission processes were occurring and were able to detect just a few of them. Much work on neutrinos has subsequently been carried out, and today this field is a very active one. Neutrinos coming from outer space have sometimes been detected experimentally by means of large tanks of chlorinated hydrocarbons placed in a deep mine, about 3 km (nearly 2 mi) deep. The neutrinos interact with chlorine atoms ($^{37}Cl$) to produce radioactive argon atoms, the number of which can be measured from the rate of their radioactive decay. The detectors sometimes present an area of two hundred square meters (twenty-two hundred square feet) and may weigh thousands of tons.

The reason that it is so difficult to detect neutrinos is that they interact so weakly with matter. The average distances they travel before striking a nucleus—such a distance is known as the *mean free path*—are enormously long. A neutron passing through ten centimeters of lead has a good chance of striking a nucleus and being scattered, but with a neutrino the situation is enormously different. The mean free path for a neutrino passing through lead is of the order of one light-year ($9.46 \times 10^{12}$ km, or 5,900 billion mi). In other words, a neutrino has a good chance of traveling about 10 trillion km (6 trillion mi) through a gigantic block of lead without interacting with any of the nuclei; that distance is more than sixty thousand times the distance from Earth to the Sun. Thus the vast majority of the neutrinos coming to us from the Sun never strike a nucleus, even if they pass right through Earth. It can be calculated that the chance of a neutrino being scattered as it passes right through Earth is only about one in a million. It is estimated that every second about 70 billion ($7 \times 10^{10}$) neutrinos pass through every square centimeter of our bodies. Although such vast numbers of neutrinos are passing through our bodies all the time, in an average lifetime only about one neutrino will have struck one of our nuclei, with no effect whatever.

These facts show how difficult it is to detect neutrinos. Until a few years ago the total number of them that had been detected was smaller than the number of scientific papers that had been written about them; it is still not much more.

In February 1987 a supernova was observed in the Large Magellanic Cloud, and it is estimated that some $10^{58}$ neutrinos were liberated. They were released 160,000 years ago, and since their mass is so tiny they travel at not much less than the speed of light. After traveling a distance of 160,000 light-years, about $3 \times 10^{14}$ of them should have passed through the large underground tanks in use for their detection. Of these, just nineteen were detected in Japan, and eight in the United States.

The earlier measurements on the neutrinos emitted by the Sun led to a difficulty. Theories of the nuclear processes occurring in the Sun allowed estimates to be made of the rate with which neutrinos would be detected by the instruments set up on Earth. However, the results obtained with chlorinated hydrocarbon detectors always showed that the number that could be detected was only about one-third of the number expected. This became known as the *solar neutrino problem.*

Recent work has resolved this difficulty, which resulted from the fact that there are three varieties, or "flavors," of neutrino and that they change their form as they travel to Earth. Since each of them has its corresponding antineutrino, there are thus six types of neutrino, designated as follows:

| | |
|---|---|
| electron-neutrinos, $\nu_e$ | electron-antineutrinos, $\bar{\nu}_e$ |
| mu-neutrinos, $\nu_\mu$ | mu-antineutrinos, $\bar{\nu}_\mu$ |
| tau-neutrinos, $\nu_\tau$ | tau-antineutrinos, $\bar{\nu}_\tau$. |

The neutrino that is involved when a neutron is converted into a proton, a process we discussed earlier, is now called an electron-antineutrino. It was originally called a neutrino, but the convention is that the less prevalent particle is always referred to as the antiparticle. The electron-antineutrino that is produced when the neutron is converted into a proton will be at once annihilated when it meets an electron-neutrino, so that the process can be written as

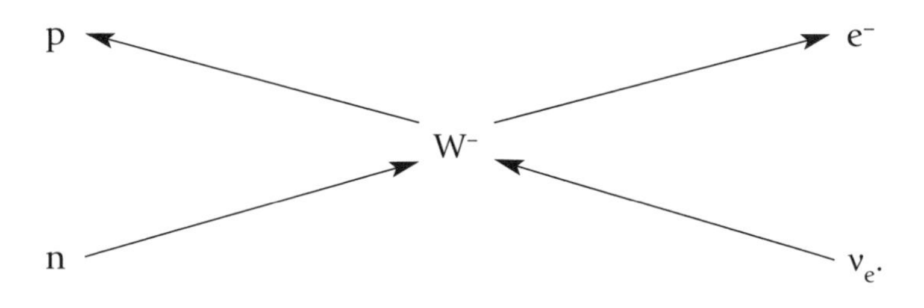

This shows the $W^-$ particle reacting with an electron-neutrino to produce an electron, instead of emitting an antineutrino as it does so; the alternative processes are essentially the same.

The mu-neutrino ($\nu_\mu$) and the mu-antineutrino ($\bar{\nu}_\mu$) have of course a similar relationship to the muons, as have the tau-neutrino ($\bar{\nu}_\tau$) and its antiparticle ($\bar{\nu}_\tau$) to the taus. We now know that neutrinos are emitted by the Sun as electron-neutrinos ($\nu_e$) but that as they travel they may be converted into mu-neutrinos and tau-neutrinos. The difficulty with the earlier work is that it was detecting only the electron-neutrinos.

In April 2002 the announcement was made of the results of important neutrino experiments carried out by a joint US-Canadian-British team of scientists at the Sudbury Neutrino Observatory (SNO), in Sudbury, Ontario. The detector, set up at the bottom of a mine 2,039 meters (6,690 feet) below the surface, was a spherical vessel containing one thousand tons of ultrapure heavy water ($D_2O$), surrounded by ultrapure ordinary water, which screens out radiation from other sources. Detection of the neutrinos is achieved by making use of two processes. One of these, which occurs only with electron-neutrinos, is

$$\nu_e + \text{deuterium} \rightarrow 2 \text{ protons} + \text{electron.}$$

The electron reacts with heavy water with the production of a photon, which is detected by photocells supported throughout the ordinary water that surrounds the heavy water. From the number of photons produced, the number of electron-neutrinos can therefore be estimated.

In addition, a neutrino in any of its three forms, denoted by $\nu_x$, interacts with a deuterium nucleus to form a neutron that adds on to another deuterium nucleus to give a tritium nucleus with the emission of a gamma ($\gamma$-) ray:

$$\nu_x + \text{deuterium} \rightarrow \text{proton} + \text{neutron;}$$
$$\text{proton} + \text{deuterium} \rightarrow \text{tritium} + \gamma\text{-photon.}$$

The number of tritium nuclei and $\gamma$-photons produced therefore provides a measure of the total number of neutrinos.

In the Sudbury experiments, over a period of 308 days reliable evidence was obtained for the detection of about 2,800 neutrinos. By detection of the radiation emitted, and in other ways, it proved possible to confirm that the rate of emission of neutrinos from the Sun is consistent with current theories of the processes occurring in the Sun. In other words, the so-called solar neutrino problem is no longer a problem. These experiments also showed convincingly that the neutrino has a nonzero mass, although an extremely small one. Its mass cannot be more than a millionth of the mass of an electron, which itself is one of the lightest of the fundamental particles.

The SNO experiments completely confirm the theories of the processes occurring in the Sun. We can now say with some confidence that we do know just how the Sun shines.

▲▲▲

During the 1950s physicists discovered many new particles and were beginning to realize that it was important to classify them and perhaps to explain them in terms of more fundamental particles. After all, the hundred or so nuclei that corresponded to the chemical elements had been found to be made up of just three particles: the electron, the proton, and the neutron. Could not these new particles be composed of a fairly small number of fundamental building blocks?

In 1961 an important classification system was proposed by American physicist Murray Gell-Mann (b. 1929). One of his contributions was to suggest that there is a new property that particles seem to have, which he called "strangeness." This sounds rather esoteric, but strangeness is really no more mysterious than electric charge, to which we have become accustomed. We are used to the idea that fundamental particles can have three possible charges: +1, −1, and 0. All the experiments that have been done on particles suggest that in any process the electric charge is conserved; a charge cannot disappear or be created from nothing. A neutron can only turn into a nucleus by a process in which a particle is produced to balance the charge, and that particle is the electron. The idea of strangeness was introduced in order to explain why there were certain other restrictions as to which processes could possibly occur. Just as electric charge must be conserved in any process, so must strangeness always remain the same. By arguing in this way Gell-Mann was able to classify the new particles in terms of their charge, spin, strangeness, and other properties, and somewhat similar ideas were put forward by Israeli physicist Yuval Ne'eman (b. 1925). An important consequence of their work was that they found a gap in their classification and predicted that there must be another particle, which they called the omega minus ($\Omega^-$), which they said would have a strangeness of –3 and a mass of 1680 MeV/$c^2$, which is more than 1.5 times the mass of the proton. This particle was discovered experimentally in 1964 at the Brookhaven National Laboratory, a great triumph for the theory. It was found to have a charge of –1, a spin of ½ (meaning that it is a fermion), and a lifetime of $10^{-10}$ seconds, which is quite long for a particle of this kind.

As an offshoot of this classification, in 1964 Gell-Mann and George Zweig (b. 1937), then still a graduate student at Caltech, put forward the idea that the hadrons are made up of more fundamental particles, which have charges that are a fraction of the charge of the electron. At the time this was a revolutionary idea, since it had always been assumed that there could be nothing smaller than the electron. Gell-Mann called these particles *quarks,* taking the name from a line in James Joyce's *Finnegans Wake*: "Three quarks for Muster Mark." This seems to suggest, incidentally, that *quark* should be pronounced so as to rhyme with *mark* rather than with *quart.* We must not think of hadrons, such as protons, as somehow having quarks rattling round inside them—any more than we should think of nuclei as having electrons inside them. The importance of the concept of quarks is that it gives us a neat classification and understanding of the behavior of the hadrons. For his contributions in this field Gell-Mann was awarded the 1969 Nobel Prize for physics.

Some of the quarks and their characteristics are summarized in Table 4. Protons, neutrons, and pions can be understood in terms of just two types of quarks, which are given the labels *up* and *down.* The

**Table 4. Quarks**

| *Flavor* | *Charge* | *Mass, in MeV/c²* | *Strangeness* |
|---|---|---|---|
| Up, u | +⅔ | 5 | 0 |
| Down, d | –⅓ | 10 | 0 |
| Charm, c | +⅔ | — | 0 |
| Strange, s | –⅓ | 200 | 1 |
| Top, t | +⅔ | — | 0 |
| Bottom, b | –⅓ | — | 0 |

*Note:* Each of these six types can exist in three "colors," red, blue, and green, so there are eighteen types of quark. For each there is a corresponding antiquark, making thirty-six types in all; in other words, eighteen have a spin of ½, and eighteen of –½.

The proton is u + u + d, with a charge of ⅔ + ⅔ – ⅓ = +1 and a spin of +½ or –½.

The neutron is u + d + d, with a charge of ⅔ – ⅓ – ⅓ = 0 and a spin of +½ or –½.

The pion $\pi^+$ is u + anti-d, with a charge of ⅔ + ⅓ = 1.

The pion $\pi^-$ is d + anti-u, with a charge of –⅓ – ⅔ = –1.

The pion $\pi^\circ$ is u + anti-u, with a charge of ⅔ – ⅔ = 0.

The omega minus, $\Omega^-$, is s + s + s, with a charge of 3(–⅓) = –1 and a strangeness of –3.

labels have no significance and are quite arbitrary. The up-quark has a charge of +⅔, and the down-quark a charge of –⅓. Each of the quarks has its antiquark, having the opposite charge. A proton is regarded as made up of two up-quarks and one down-quark, giving it a charge of 1. A neutron has one up-quark and two down-quarks, which means that it has a charge of zero. The pions, with their smaller mass, consist of two quarks.

▲▲▲

For several decades physicists have been striving to develop theories that give a comprehensive and unified explanation of the four fundamental forces of nature. This has proved a difficult task, and so far it has been impossible to find a theory that includes gravity—the force that has been known the longest—along with the other three forces. It is convenient to list first some of the theories that have had some success, with a brief explanation, and then to discuss them in a little more detail.

In roughly increasing order of comprehensiveness and complexity, some of the theories are as follows:

1. Quantum electrodynamics (QED), concerned only with electromagnetism.
2. Electroweak theory, concerned with the electromagnetic force and also the weak force.
3. Strong-force theory, concerned with the force within atomic nuclei.
4. Quantum chromodynamics (QCD), also concerned with the strong force operating within nuclei, dealing specifically with how quarks interact with one another.
5. Grand unified theories (GUTs), which attempt to combine QED with QCD and with the theory of the weak nuclear force as well. It does not include gravity.
6. Theory of everything (TOE), which includes gravity in addition to the other three forces. Not much progress on this has been made so far.

Quantum electrodynamics is a highly successful theory that has been tested experimentally a large number of times and has passed every test. The theory is a development in modern terms of Maxwell's electromagnetic theory and has its roots in a suggestion made in 1932 by Hans Bethe (b. 1906) and Enrico Fermi that electromagnetic forces are related to the exchange of photons between particles. Later, in the 1960s, the theory was developed mathematically by American physicist Julian Seymour Schwinger (1918–94) and independently by Japanese physicist Sin-Itiro Tomonaga (1906–79). At about the same time but independently Richard Feynman (1918– 88) developed QED in a completely different and particularly clear way. All three shared the 1965 Nobel Prize in physics for this work. Bethe was awarded the 1967 prize, but that was mainly for his work on nuclear processes in the stars.

Since the photons that relate to the electromagnetic forces have zero mass, they travel at the speed of light and, as a result, the electromagnetic force is a long-range force. According to the electroweak theories, on the other hand, the weak force is related to particles that have mass. These are the intermediate bosons, the $W^+$, $W^-$, and Z particles, which we discussed earlier. The weak-interaction theory is a little more complicated than the theory of electromagnetic radiation, because instead of one particle, the photon, three particles are involved. The theory of the electroweak force was developed independently by Sheldon Lee Glashow (b. 1932), Abdus Salam (1926–96), and Steven Weinberg (b. 1933). All three shared the 1979 Nobel Prize in physics for this work.

According to their theory, these intermediate bosons are created out of nothing, using borrowed energy, and they are referred to as *virtual particles*. This curious process is allowed by the uncertainty principle, but only on the condition that the energy is repaid very quickly. The bigger the mass, the more quickly the debt has to be repaid. It is for this reason that whereas the electromagnetic force involves the photons—which, having no mass, can survive longer and travel farther—the carriers of the weak force, having mass, are short-lived and can travel only a short distance before they are annihilated. The weak force is therefore short-ranged.

According to strong-force theory, the protons and neutrons in a nucleus are held together by the exchange of pions. We met the pions earlier, in connection with the theoretical work of Yukawa and their discovery by C. F. Powell. Again, the fact that the strong force is a short-range force, extending only about the diameter of the nucleus, follows from the fact that the pions are particles having mass, so when they are created they have a very short life. The reason that the nuclei are so tiny is that the strong force has such a short range.

The shortness of the range of the strong forces is of great importance, since otherwise the forces would pull all of matter together, crushing ordinary atoms out of existence. Yukawa was the first to realize that if the virtual particles, the pions, had a certain mass they could live just long enough to carry the strong force between adjacent nucleons in a nucleus but not long enough to reach out as far as neighboring nuclei. He predicted that his hypothetical particle would have a mass of about 200 times that of the electron, namely about 100 MeV/$c^2$. The pion actually has a mass of 140 MeV/$c^2$.

According to modern theory, individual protons and neutrons lose their identity in the nucleus, which should be regarded as composed of quarks bonded together by forces similar to the strong forces. This is the subject of the field of quantum chromodynamics (QCD). This expression was deliberately chosen to be similar to quantum electrodynamics (QED), according to which charged particles interact with one another by the exchange of photons. The theory of QCD is based on the postulate that the bonding together of the quarks in the nucleus occurs by the exchange of particles that are called *gluons*—because they glue the quarks together. The prefix *chromo-* relates to the fact that quarks and gluons are supposed to come in different "colors"—not, of course, in the sense we understand color.

The expression *grand unified theory* (GUT) is somewhat too grandiose, because no GUT attempts to include gravity, which has so far proved elusive. What a GUT does is to try to find a comprehensive theory that explains electromagnetic interactions as well as the strong and weak forces. Several such theories have been put forward, but there is still no general agreement about the validity of any one of them. Some scientists in fact think that a grand unified theory will never be found.

The same applies to the attempts to include gravity in addition, in a theory of everything (TOE). One theory that is being developed to do this is *string theory*. The term is used to describe a theory that attempts to describe the fundamental particles and the four forces of nature in terms of tiny entities that are one-dimensional; in other words, they are strings having length but no thickness. These strings are very much smaller than the smallest observed particles, electrons. Sometimes they form loops, but even these are much smaller than electrons. Strings are in fact supposed to be of the order of magnitude of what is known as the *Planck length*, which is about $10^{-35}$ meters, some $10^{20}$ times smaller than an atomic nucleus.

This is an active field of research, and several formulations of string theory have been put forward. Michael Green (b. 1946), now a professor of theoretical physics at Cambridge, has played a pioneering and major role in the formulation of string theories. In some of the theories it is postulated that there are more than the usual three dimensions of space. It must be admitted that at present there is no clear-cut scientific evidence that favors string theories, and some prominent scientists are of the opinion that they will never be successful. However, the only thing we should ever predict is that the unexpected may well happen.

▲▲▲

The story of the elementary particles is somewhat complicated, and much more theoretical and experimental work remains to be done. Now that we have been introduced to the essential elementary particles we can summarize the whole situation quite briefly. According to what has come to be called the *standard model*, all of matter is supposed to be made up of a limited number of leptons and quarks, which are the truly fundamental particles. They are said to occur in three *generations*, or families, as shown in Table 5. The first generation consists of the electron and its neutrino, which are leptons, and the up and down quarks, which are hadrons. The second generation consists of the muon (a heavy electron) and the corresponding neutrino and quarks, which are heavier than their first-generation counterparts but otherwise have similar properties. The third generation consists of a still heavier set of particles, the tau particle and its corresponding neutrino and quarks.

The hadrons, the chief examples of which are the protons and neutrons, are the particles that involve the strong force, and they are made up of various combinations of quarks.

All known particles can be accounted for within this classification, which has led to the prediction of the existence of many other particles that have later been confirmed by experiment.

**Table 5. The three generations of particles**

| *Generation 1* | *Generation 2* | *Generation 3* |
|---|---|---|
| Electron, e | Muon, $\mu$ | Tau, $\tau$ |
| Electron-neutrino, $\nu_e$ | Muon-neutrino, $\nu_\mu$ | Tau-neutrino, $\nu_\tau$ |
| Up-quark, u | Charm-quark, c | Top-quark, t |
| Down-quark, d | Strange-quark, s | Bottom-quark, b |

7

# OUR WORLD AND OUR UNIVERSE

We can get some feeling for the physical world around us by first taking a brief look at the history of geology, which was one of the latest sciences to develop. Little of any significance was done in geology until the last decade of the eighteenth century, although a few interesting geological observations had been made earlier. Leonardo da Vinci (1452–1519), for example, during his work as a young man on the construction of canals in Italy, examined the rocks and saw many marine fossils in them. He correctly deduced that the mountains of northern Italy had at one time been covered by the sea. British scientist Robert Hooke (1635–1703), a contemporary and bitter rival of Newton, also observed fossils and concluded from the thickness of the beds in which they lay that they must have been formed much earlier than had been popularly supposed. Many people then had strong preconceptions about the subject, based to a great extent on a literal interpretation of Scripture, particularly the book of Genesis. Consequently, deductions from geological evidence that should have been convincing were often rejected on the basis of these prejudices and not for any scientific reason.

One of the leading early geologists was Abraham Gottlob Werner (1749–1817), professor at the mining academy in Freiburg, Germany. He was one of the first to realize that the Earth's crust does not consist of a chaotic jumble of rocks but an ordered succession of layers, the oldest at the bottom and the youngest at the top. He became the chief exponent of the so-called *Neptunian theory*, according to which the mountains and valleys of Earth always had the same general form as today but had been submerged under water, which contained the substances present in today's crust. By a complex series of events these substances became deposited, and the excess water eventually disappeared. There were a number of difficulties with this theory, one being that volcanoes could not be explained without a completely different type of theory. The opponents of the Neptunists were the *Vulcanists*, who ascribed a fiery origin to certain rocks. Most geologists at the time accepted the idea of *catastrophism*, believing that major catastrophes, like the biblical flood, were necessary to explain Earth's development.

Scottish geologist James Hutton (1726–97; see fig. 47) made contributions of the greatest importance in this area. His career had a rather disorganized and unpromising beginning. He was born in Edinburgh and after leaving school was first an apprentice in a

**Figure 47.** Scottish geologist James Hutton (1726–97) is regarded as the founder of modern geology. He established that Earth's internal heat has much to do with rock formation. He was the first to abandon older ideas about geological changes, replacing them by his theory of uniformitarianism, according to which changes take place continuously over vast periods of time. (Edgar Fahs Smith Collection, University of Pennsylvania Library)

lawyer's office. He then went abroad and studied medicine at Paris and Leiden, qualifying as an MD at the University of Leiden. He never practiced medicine but turned to farming, studying agriculture in England and elsewhere, and returned to a family farm near Edinburgh. He then became fascinated by chemistry and more particularly by geology. After making a fortune out of inventing a method of manufacturing the chemical sal ammoniac (ammonium chloride) he settled in Edinburgh in 1768, devoting the rest of his life to geological and other scientific investigations.

In Edinburgh he numbered among his friends several distinguished men, such as chemist Joseph Black, philosopher David Hume, economist Adam Smith, engineer James Watt, and painter Henry Raeburn, who painted his portrait. Hutton made many geological observations of his own, mainly in Scotland, and formulated general principles that are now widely accepted. Up to that time the biblical flood was regarded as playing the decisive role in shaping Earth, but Hutton realized that this could not be the case, even if the flood had been of long duration. He proposed an alternative theory, known as *Plutonism*, or *uniformitarianism*, according to which geological events occur in a continuous manner rather than as a result of catastrophic events such as floods. His theory placed special emphasis on the role of rivers in excavating valleys and in depositing dissolved and suspended material. Sediment carried along by rivers would be washed into the sea and accumulate in deposits, which might form new rocks by the action of heat coming from the interior of the planet.

Perhaps under the influence of his friend James Watt, Hutton regarded Earth as a kind of gigantic steam engine, its subterranean heat creating upheavals from time to time. He did not consider that catastrophic events such as floods play any major role. He wrote of "a cyclic progression of changes so ancient as to obscure any vestige of a beginning and no prospect of an end." In 1785 Hutton presented his ideas in a book, *Theory of the Earth, with Proofs and Illustrations*, which was expanded over the next decade. At first there was much opposition to Hutton's theory, but he is now commonly regarded as the founder of the modern science of geology. The three decades in which his ideas were having the greatest influence, 1790–1820, have been termed the "heroic age" of geology.

Another important geologist was Jean Louis Rodolphe Agassiz (1807–73), of Switzerland, who emphasized the role played by glaciers in forming Earth's surface. He concluded that much of northern Europe had at one time been covered in ice; evidence has since been obtained that there have been several ice ages. Much support was given to the ideas of both Hutton and Agassiz by Charles Lyell (1797–1875), who was born the year that Hutton died. He was not a highly original geologist but made an important contribution with his *Principles of Geology*, published in three volumes from 1830 to 1833. It is an interesting commentary on the times that this book was widely read by the educated public, had larger sales than most books by popular novelists, and ran to several editions. The book followed Hutton's ideas for the most part, but there were small but significant differences in point of view. For example, Hutton's theory was that few geologic changes were still taking place, but Lyell showed by his direct observations that the coast of Sweden was still rising at a significant rate. It was thus possible for mountains to be formed gradually over long periods of time, rather than by any catastrophic convulsion.

Of all the essentially correct geological theories, *continental drift* was one of the most persistently opposed.[1] The idea behind the theory is that the continents have changed their positions over time. Several geologists suggested this, but the first to build a convincing case was Alfred Lothar Wegener (1880–1930), often regarded as the father of the idea of continental drift. He received his doctorate from Heidelberg in astronomy rather than geology and was primarily a meteorologist. On the basis of several lines of evidence, including the distribution of fossils, he proposed that the continents may once have been joined up into one supercontinent, which he called Pangaea. This broke up about 200 million years ago, the fragments drifting apart to form the continents as they exist today. He first published this idea in a book in 1912, with various later editions, and in 1924 there was an English translation, *The Origin of Continents and Oceans*.

In North America this theory was at first treated with indifference, whereas many European scientists reacted with hostility. The main objection to the theory was that there seemed to be no obvious driving mechanism for the movement of the continents. Unfortunately, Wegener had been too specific in suggesting a rate at which the continents spread, and when appropriate observations were made, his rate was shown to be wrong. As often happens with new theories in science, the proverbial baby was thrown out with the bathwater—that is to say, the whole theory was thrown aside by many geologists because of incorrect details. The same thing happened to Darwin with his theory of evolution (see chap. 9): tiny points on which he went wrong (and there were surprisingly few of them) were taken to be evidence that the whole idea was wrong.

The evidence adduced by Wegener and others in favor of continental drift now seems so convincing that it is hard to see why there were so many objections. Rock formations in western Africa, for example, can be seen to be clearly matched with ones in Brazil. After decades of hostility to the idea some progress on this subject was made, partly as a result of the work of British geologist and geophysicist Arthur Holmes (1890–1965), a pioneer in the use of radioactive decay methods for the dating of rocks. In 1928 he suggested that convection currents within Earth's mantle might provide the driving mechanism for continental drift. At first little attention was paid to his work, but in 1944 his influential book *The Principles of Geology* appeared, in which he convincingly put forward his theories.

Although Earth is our home, it is not easy to obtain information about it. Its radius is about 6,400 km (about 4,000 mi), but the deepest that a person has ever reached is a mere 4 km (about 2.5 mi), and the deepest borehole ever drilled reached less than 20 km (about 12 mi) below the surface. We therefore have to get our information about Earth's interior by indirect means. Much of what we know comes from the study of seismic waves, which are the waves resulting from earthquakes or other shocks. They travel through rock and are reflected and refracted by boundaries between different layers of rock, much as light is reflected and refracted from the surface of a block of glass. A number of seismic stations have been established at different parts of the world, and by monitoring the seismic waves it is possible to infer something about how Earth is constructed.

Some of the conclusions about Earth's structure are shown in figure 48. Earth has a solid inner core with a radius of about 1,500 km (950 mi). This is surrounded by a molten outer core about 1,900 km (1,200 mi) thick, so that the total radius of the core is about 3,400 km (2,150 mi). The core is mainly composed of iron. The temperature of the inner core is about 5,000°C (9,000°F), which means that it is literally red hot, and the pressure is about 3 million times normal atmospheric pressure. The pressure in the outer core is lower, which is why the material is liquid rather than solid. Above the core is the lower mantle, considerably cooler and about 2,900 km (1,400 mi in thickness and solid in its deepest regions, because of the high pressure. This solid region is also called the *mesosphere* (from Greek *meso*, 'middle'), a word that is also confusingly applied to one of the layers of Earth's atmosphere.

Above the mesosphere but still a part of the mantle is what is called the *asthenosphere*, from Greek *asthenes*, 'weak.' The reason for the name is that this is a region of weakness, the rocks being partially melted; this is because of the composition, temperature, and pressure of this region. On account of this partial melting, seismic waves travel more slowly than they do above and below this zone.

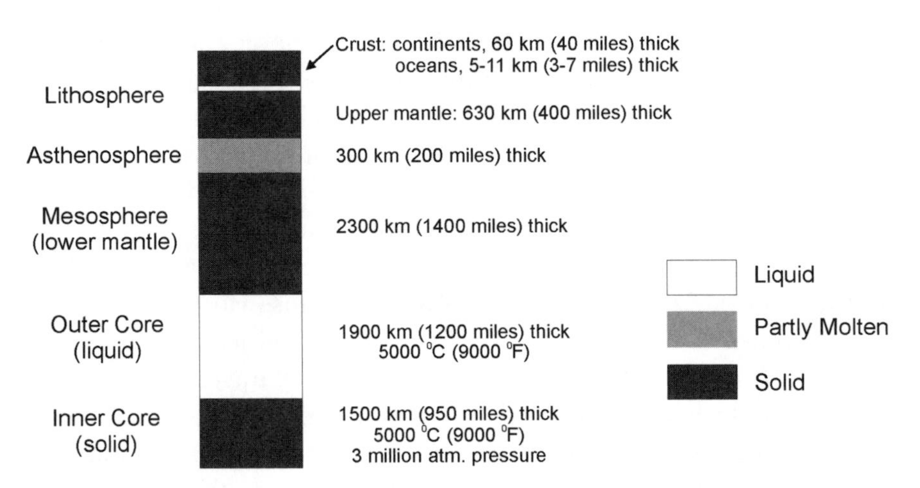


**Figure 48.** Earth's layers, showing the approximate depths at which the boundaries between them exist.

Everything above the asthenosphere is called the *lithosphere*. Because of the slushiness of the asthenosphere the material above it is floating on it. The continents can therefore move about slowly on the surface, gradually changing the geography of the planet. The upper part of the lithosphere is called Earth's crust, which in the land areas is up to 60 km (40 mi) thick, and under the oceans only 5–11 km (3–7 mi) thick.

The fact that the ocean crust is relatively thinner than the continental crust was discovered in the 1950s, when the seismic technique began to be used far out to sea, the waves sometimes being produced by underwater explosions. At the same time the oceanic floor was found to be rugged, with mountains and canyons as well as great ridges and valleys sometimes thousands of kilometers in length. Important discoveries in the field were made by American geophysicist Harry Hammond Hess (1906–69), who spent most of his career at Princeton University. In 1962, following the discovery of the extent of the midocean ridges and valleys, Hess proposed the hypothesis of sea-floor spreading. He suggested that material was constantly rising from the mantle to form the midocean ridges, which then spread out to form a new oceanic crust. He suggested that this process would continue as far as the continents, where the oceanic crust would sink beneath the continental crust.

Another important contribution was made by Edward Crisp Bullard (1907–80), professor of geodesy and geophysics at Cambridge University. He was the first to use computers to track the movement of land masses and in 1964 presented a detailed and convincing analysis of the data. After that the idea of continental drift became generally accepted, and this branch of science was called *plate tectonics*. Canadian geophysicist John Tuzo Wilson (1908–93) was initially a staunch opponent of the idea of continental drift, but his later work provided much support for it. It was he who in 1965 suggested the concept of the *transform fault*, which occurs when plates in the crust slide past one another, rather than one of them sinking below the other. This newer version of the theory deals in more detail with the formation, movement, and destruction of Earth's outer shell, which includes both the continental and oceanic crust. The basic idea is that the lithosphere, formed by the solid crust, consists of large, tightly fitting plates that float on the semimolten layer of the mantle beneath them. The plates move relative to one another, and over the course of geological time this movement has brought the landmasses into the present arrangement of continents and oceans. Continents drift at about the same rate as the growth of our fingernails.

▲▲▲

Having looked briefly at Earth's structure we should now consider what the rest of the universe is like. The science that deals with the universe is astronomy, which is the oldest of the numerical sciences. Research in this field began in about 3000 BCE, when people, particularly in the Near East and China, began to group the stars into constellations. Monuments such as the pyramids in Egypt and Stonehenge in England provide evidence for an early interest in astronomy. The relative movements of the Sun and the Moon led to a primitive form of calendar, in which the year, the month, and the week were defined. The first satisfactory proof that Earth is spherical was given by Aristotle (384–322 BCE), who would later create much controversy by his insistence that Earth was the center of the universe. Aristarchus (c. 320–c. 250 BCE), the most accurate and original of the Greek astronomers, attempted to estimate the distances from Earth to the Sun and the Moon, but his observations were rather inaccurate by modern standards. He was the first to propose that it is better to regard Earth as going around the Sun, rather than the other way round, as was then generally assumed. Another famous Greek astronomer was Hipparchus of Rhodes (c. 170–c. 125 BCE), who constructed a detailed catalog of 850 stars and suggested a scale of magnitudes to indicate their relative luminosity.

The Egyptian-Greek astronomer Claudius Ptolemaeus (usually known as Ptolemy, c. 90–170 CE) made a detailed study of the movement of the Sun and the planets and proposed a detailed mathematical model in which Earth was at the center and the Sun and planets moved around it in complicated orbits. This Ptolemaic universe was accepted by the Western world for about thirteen centuries, during which little progress was made in astronomy or indeed in any branch of science. The main reason for this is that for many centuries scientific procedures became what we today call "politically incorrect" and were supplanted by theological and astrological studies, which are much less taxing to the human brain.

The first step out of the intellectual darkness was taken by Polish priest Nicolaus Copernicus (1473–1543), working mainly from the astronomical data of others rather than from his own observations. Copernicus recognized that in explaining the movements of the planets it was more satisfactory to assume that they revolve around the Sun rather than in the manner assumed by Ptolemy. He first expressed his ideas privately in 1514 in a short manuscript and continued to develop them for the next thirty years. He explained the apparent movement of the Sun around Earth as due to Earth's rotation. He also suggested that the stars are much farther away than had previously been proposed. He expounded his ideas in a book *De Revolutionibus Orbium Coelestium* (The Revolution of the Heavenly Spheres), which was published in the year of his death; he may have seen a copy of the published book only on the day he died. Copernicus's work was one of the most important and original advances in astronomy.

Much more astronomical information was required to support Copernicus's ideas. Important work was done by Danish astronomer Tycho Brahe (1546–1661), who greatly improved the accuracy of astronomical observations and made a vast number of observations, measuring the positions of 777 stars with great accuracy. Brahe never became a Copernican but took the compromise position that all of the planets except Earth go around the Sun, which itself goes around Earth. The Sun does, of course, go around Earth; they go around each other. What Copernicus had pointed out was that we get a mathematically simpler concept of the entire solar system by regarding all the planets as going around the Sun. It took many years to abandon Brahe's compromise, which led to predictions that were very similar to those of Copernicus.

In 1600 German astronomer Johannes Kepler (1571–1630) went to work with Brahe, and there is a story that on his deathbed Brahe begged Kepler not to use his observations to support the Copernican hypothesis. Kepler's great contribution was to discover that planets have elliptical rather than circular orbits and to establish empirical laws describing their motion. His work did support Copernican ideas, but Kepler placed more stress on the shapes of the orbits rather than on whether Earth or the Sun was the center of the system.

Galileo Galilei (1564–1642; see fig. 49) developed a greatly improved telescope, with which he was able to make a series of spectacular discoveries. In 1609–10 he made observations of four satellites of Jupiter and of the orbit of Venus. All of his work led him to accept the Copernican model for the solar system, with the Sun at its center. It is well known that the Roman Catholic church ordered Galileo to deny the Copernican theory. An interesting detail is that Galileo sold his telescope to the Venetian Republic for commercial and military use; this is an early example of the connection that often exists between pure science and practical applications.

Galileo also made important contributions to the science of mechanics. In particular he established by sliding weights down

**Figure 49.** Galileo Galilei (1564–1642) made significant contributions to both physics and astronomy. He was the first to understand and to express mathematically the motion of falling and sliding bodies. He made many important astronomical observations and obtained strong evidence in favor of the Copernican theory of the motions of the planets. (Edgar Fahs Smith Collection, University of Pennsylvania Library)

inclined planes that their motion was independent of their mass; previously it had been assumed that a heavy weight would fall faster than a light one. The story that he dropped weights from the top of the Leaning Tower of Pisa is a nice one, but unfortunately it appears to be untrue.[2] In other experiments Galileo established the relationship between the force on a body and its acceleration. However, he was not able to relate his mechanics to his astronomical observations on the orbits of planets; that was done by Newton.

Francis Bacon (1561–1626), who was almost contemporary with Galileo, emphasized that general principles must always be tested against the experimental results. His writings exerted a great influence, and in his most significant work Newton followed the path laid out by Bacon. Isaac Newton was born in the year of Galileo's death and was still a young man at Cambridge when he clarified Galileo's mechanical principles, stating his famous three laws of motion. Even before that he had stated his law of gravitation. The story is that a falling apple in an orchard made him think in terms of a force of attraction between Earth and the apple and that he then went on to see if the motion of the planets could be explained in terms of such a force. One matter that troubled him for many years in formulating his theory was whether it was correct to regard the mass of a body such as a sphere as if it all resided at the center of the sphere. It is correct, but it is not easy to prove that it is. Newton's law of gravitation stated that the force of attraction between two spheres is the product of the masses and is inversely proportional to the square of the distance between the centers of the spheres. That means that if there is a certain attraction when two bodies are a certain distance apart, if we were to double the distance, the force would decrease by a factor of two squared, which is four. This *inverse square law* applies to many things beside gravity. We have seen that it applies to the force between two charged bodies. It applies in a rough way to everyday situations, such as how clearly we can see objects and how well we can hear sound (aside from complications due to reflection of sound).

Newton's great contribution to science was to unite celestial and terrestrial science for the first time by interpreting the motion of the planets in terms of his gravitational law and of his mechanical laws. All of his studies gave great support to the Copernican theory. His detailed mathematical work on the motion of planets was followed by a similar treatment by Edmund Halley (1656–1742) on the famous comet known by his name. This comet approached Earth in 1682, and Halley correctly predicted that it would do so again in 1758.

Newton's mathematical treatment of the solar system was one of the greatest achievements made in astronomy, and it led the way to similar investigations into outer space. Much progress was made by William Herschel (1738–1822), who applied Newton's theory to binary stars, which are systems of two stars orbiting each other. Another of Herschel's many important contributions was to study the Milky Way in detail. Astronomers were at that time just beginning to understand the significance of that remarkable band of light that stretches across the night sky. When Herschel did his work, the details of the Milky Way were unknown, but the suggestion had been made by English instrument maker Thomas Wright (1711–86) that the Milky Way is part of our galaxy, which has the shape of a flat disk; this is now known to be correct. Herschel constructed several telescopes, each larger than the last. He observed a number of objects called *nebulae*, which appear different from the stars in the Milky Way. He found that when he increased the size of his telescopes he could distinguish individual stars in some of them. He correctly thought, but could not prove, that many of these nebulae were star systems like the Milky Way, lying outside our galaxy. Later these came to be known as *extragalactic nebulae*. Some of them, including our own, have spiral structures and are known as *spiral nebulae*. Today all such nebulae are usually called *galaxies*.

A problem for Herschel and all other astronomers in his time was that they could not measure distances from Earth to even the nearest stars, let alone the nebulae. It is important to do this, and first we should consider a convenient unit for measuring distances, the *light-year*, which is the distance that light travels in a year. Since light travels at about 300 million meters a second, or about 300,000 km (186,400 mi) a second, and a year is about 32 million seconds long, a light-year is nearly 10,000,000,000,000 (10 million million, or $10^{13}$) km. The following may help us to appreciate these numbers. Light takes about a hundredth of a second to cross the Atlantic. It takes 1.3 seconds for it to get to the Moon, which on the average is about 384,400 km (238,870 mi) from Earth, so we can say that the Moon is roughly 1.3 light-seconds away from us. Light reaches the Sun in about 8 minutes, so the Sun is 8 light-minutes from us. All of the stars that we see in the sky are more than 4 light-years away, and some are several billions of light-years from us. Astronomers, by the way, usually measure distances not in light-years but in parsecs; one parsec is equal to 3.262 light-years, or $3.086 \times 10^{13}$ (about 30 million million) km (about 18.6 million million mi).

The nearest star, Proxima Centauri, is now known to be about 4.3 light-years away from us. It is difficult for us to appreciate such great distances, and the following may help. Suppose that we had a spaceship that would travel at an average speed of 10 kilometers a second, which is 36,000 km an hour (22,400 mph). That speed is about Mach 30, or thirty times the normal speed of sound. Such a speed is substantially greater than any speed yet recorded for a vehicle but may be possible in the future.[3] The speed is nevertheless only a small fraction, namely 1/30,000, of the speed of light, so to travel a light-year would still take 30,000 years. To get to Proxima Centauri would take about 130,000 years. Proxima Centauri does not seem to have any planets, and the nearest star that may have a planet is about 20 light-years away; to get to that would take about 600,000 years.

These facts have serious implications for those who think that we might one day be invaded by creatures something like ourselves (humanoids) or that we might be able to colonize other regions of the universe. It is not possible for life to be maintained on anything except a planet, since everywhere else is much too hot or much too cold. We know that the other planets in our own solar system are inhospitable to any form of life, so that we have to contemplate traveling at least 20 light-years. However, a trip of 600,000 years is surely too much to contemplate. Travel to or from a planet in any of the other galaxies at such a speed would take much longer than the present age of the universe!

The measurement of stellar distances was first accomplished successfully and almost simultaneously by German astronomer Friedrich Wilhelm Bessel (1784–1846), Scottish astronomer Thomas Henderson (1798–1844), and German-born Russian astronomer Friedrich Georg Wilhelm Struve (1793–1864). They used the parallax method, which is that employed by surveyors today: by observing an object from two different positions, its distance from us can be calculated. All of the stars and nebulae are so far away that reliable measurements cannot be made from two positions on Earth, since they are not far enough apart. Copernicus had looked for a stellar parallax using observation points a distance apart on Earth but had concluded that the stars were so distant that the parallax was immeasurably small. However, in its orbital motion around the Sun, Earth moves a considerable distance. Thus observations made at different times, when Earth is at two positions the distance between which is accurately known, provide a much longer base for parallax measurements. In this way it is possible to obtain reliable values for the distances of the stars nearest to Earth. Today much more accurate parallax measurements can be made from artificial satellites, such as the European satellite *Hipparcos*. Methods other than the parallax method must be used for all the nebulae (galaxies) and for the more distant stars.

Bessel was the first to announce, in 1838, the measurement of a star's distance by the parallax method. He made his first measurement on the binary star 61 Cygni and found its distance to be 10.3 light-years. The latest precise measurements made from the satellite *Hipparcos* are 11.36 light-years for the brighter component and 11.43 light-years for the fainter one. Bessel's value was thus somewhat too low but nevertheless a great achievement in its time.

Thomas Henderson, who was director of the Royal Observatory at the Cape of Good Hope and later Astronomer Royal for Scotland, published his parallax results just three months after Bessel's announcement. They related to the star Alpha Centauri, which is about 4.4 light-years from Earth. Henderson had actually made his measurements six years earlier, in 1832, but failed to achieve priority for this pioneering work only because of his reluctance to publish his results before he had thoroughly checked them. Wilhelm Struve, member of a remarkable family of astronomers, had a similar experience; in 1837 he made parallax measurements of the star Vega (about 26.4 light-years away) but published them only after Bessel's announcement.

In the nineteenth century a discovery was made in physics that had important consequences for astronomy. It was first made in 1842 for sound waves by Austrian physicist Christian Johann Doppler (1803–53), then a professor of mathematics at the *Realschule* (state secondary school) in Prague. If sound waves of a certain pitch or frequency are emitted by a source approaching an observer, they appear to have a higher pitch than if the source is stationary. If the source is moving away from the observer, the pitch is lower. A delightful test of the effect was made in 1845 in the countryside near Utrecht, at the suggestion of Dutch meteorologist Christoph Heinrich Dietrich Buys Ballot (1817–90). An open railway carriage containing an orchestra of trumpeters was pulled past a group of musicians. The musicians confirmed that as the train approached them the pitch was indeed higher, and that it became lower after the carriage had passed them. In these modern days of high speeds we have many opportunities of hearing the Doppler effect, at airports for example, or when police sirens pass us at high speed.

Doppler suggested that the same effect would be found with light waves. If a star is receding from us, the light it emits will be of a lower frequency than if the source were at rest. Since the low frequency end of the visible spectrum is the red end, this means that the lines in the visible spectrum of the star shift toward the red end, and as a result we often speak of *redshift.* The Doppler effect, however, has nothing to do with the color of stars. The blue light from a receding star is indeed shifted toward the red, but some of the star's ultraviolet light is shifted into the blue part of the spectrum, so that the color observed may hardly change.

Doppler himself made the error of thinking that the overall color is changed, but Buys Ballot pointed out that this is not correct. Estimates of the velocities of stars from their colors led to unreasonably large values, and it was realized that other factors, particularly the age of a star, influence its color, which therefore provides no information about its speed.

However, the Doppler effect soon became a powerful tool in astronomical spectroscopy. This was first pointed out by Armand-Hippolyte-Louis Fizeau, whom we met earlier; he was the first to accurately measure the speed of light. In the spectrum of a receding galaxy, every line has a higher frequency, which means a lower wavelength, than when we observe the spectrum of the same substance in the laboratory. The first redshift for a star was observed by British astronomer Sir William Huggins (1834–1910), who made observations of the star Sirius, which is about 8.7 light-years from us, and on a number of other stars.

Contributions of great importance were made by distinguished American astronomer Edwin Powell Hubble (1889–1953; see fig. 50), who did most of his work at the famous Mount Wilson Observatory in California. He made careful studies of some of the nebulae and made estimates of their distances from Earth, finding that they were considerably farther away than the most remote parts of the Milky Way. He confirmed that the spiral nebulae were distant galaxies, referred to as extragalactic galaxies. This was an advance of great significance for astronomy. The galaxy nearest to us, Andromeda (M31), is now known to be 2 million light-years away; this is ten times farther than the most distant stars in our own galaxy. With the most modern techniques, astronomers are now able to see some galaxies that are over 10 billion light-years away. We must, of course, bear in mind the obvious fact that the light we see from a galaxy that is 10 billion light-years away is light that was emitted 10 billion years ago; anything we can learn about such a galaxy is very ancient history. Moreover, the place at which it is now observed is where it actually was 10 billion years ago. Since the universe is about 12 billion years old, that galaxy therefore appears to be where it was about 2 billion years after the universe was formed.[4] The existence of so many galaxies at such great distances gives us strong support for the great age of the universe and for its enormous size.

More recent work in astronomy has established that a galaxy is a system of stars, gas, and dust held in one region of space by gravitational attraction. The galaxies are not spread around randomly in

**Figure 50.** American astronomer Edwin Powell Hubble (1889–1953) made contributions of outstanding importance to astronomy. He found that spiral nebulae, such as Andromeda, are independent galaxies. He discovered that galaxies recede from us with speeds that increase with their distance. He also made reliable estimates of the size and age of the universe. (Courtesy of the Archives, California Institute of Technology)

space; most are in clusters, pulled together by gravity. Our own local group, besides our own galaxy, includes the Andromeda galaxy together with twenty or so smaller galaxies. The visible part of our galaxy consists of a flat disk of stars with a diameter of about 80,000 light-years and a thickness of 6,000 light-years; it also has enormous spiral arms. Our solar system, consisting of the Sun and its planets, is about 30,000 light-years from the center of the disk. Thus, when we look out from Earth along the plane of the flat disk we see many more stars than when we look in other directions. This large group of stars constitutes the Milky Way. Recent work has shown that all galaxies are much larger than their visible regions.

During the twentieth century, scientists discovered that radio waves are also received from outer space, and the investigation of these has revealed many sources of radiation that could not be detected by optical techniques. (Refer to figure 31 on p. 106 to see just where radio waves are, relative to other parts of the spectrum.) The photons in radio waves and microwaves carry about a million times less energy than those of visible radiation. Sir Martin Ryle (1918–84), who won the 1974 Nobel Prize for physics and was Astronomer Royal for a term, had a memorable way of calling attention to the smallness of the energy. When members of the public came to his observatory outside Cambridge they were invited to pick up a piece of paper as they left. On the paper was written, "On picking this up you have expended more energy than has been received by all the world's radio telescopes since they were built." By the early seventies all the world's telescopes had not yet received enough microwave energy to strike a match. Perhaps now we should say *two* matches.

The first study of radio waves from space was made in 1933 by American radio engineer Karl G. Jansky (1905–50) of the Bell Telephone Laboratory in Holmdel, New Jersey. He was working at radio frequencies of 20.3 megahertz (MHz), which corresponds to a wavelength of about 15 meters, used in shortwave radio transmission. He detected a certain background radiation that varied with the time of day and that he proved did not come from thunderstorms or from the Sun. He finally established that they were associated with the Milky Way and that they were strongest in the center of our galaxy. This discovery made the front page of the *New York Times*, but Jansky did little further work in the field, and astonishingly his results at first were largely ignored by astronomers.

One who did pay attention to Jansky's observations was another American radio engineer, Grote Reber (1911–2002). In 1937 at his own expense he built the world's first radio telescope in his backyard and for several years was the only radio astronomer in the world. His telescope could be steered and had the paraboloid shape with which people are now familiar, its diameter being 9.6 meters. He confirmed Jansky's observation of radio waves from the center of our galaxy. He was the first to detect radio waves from the Andromeda galaxy and from the Sun. Between 1940 and 1948 Reber produced contour maps showing radio wave intensities in different parts of the sky. His work had a profound influence on the later development of the field of radio astronomy.

During World War II, in 1942, British engineer James Stanley Hey (b. 1905) was investigating what was thought to be jamming of British radar by the Germans. He found that the maximum interference appeared to follow the movement of the Sun. At that time an active sunspot was traversing the solar disk, and Hey concluded that this was responsible for the emission of the radio waves of about one meter wavelength. In a sunspot region there is emission of electrons and other charged particles moving at high speeds; when these pass through the powerful magnetic field of the Sun, radio waves are emitted. In 1946 he detected a powerfully emitting radio galaxy called Cygnus A. He also detected radar reflections from meteors and initiated procedures for tracking meteors, which could be done both day and night.

After World War II, partly as a result of surplus radar equipment and growing expertise in radio communication, many remarkable advances in radio astronomy were made. In particular, many radio interferometers were built, based on essentially the same principle involved in Thomas Young's experiments on the interference of light. This equipment was highly effective in the study of radio emission from the Sun and other stars.

There is one special and useful feature of radio waves: they pass freely through clouds, unlike radiation in or near the visible part of the spectrum. Because of the frequent cloudy skies in Britain, the amount of optical astronomical spectroscopy that could be done was greatly limited, and the sunnier United States seized the initiative, erecting many observatories particularly in the west and south. Radio astronomy, on the other hand, could be carried out in Britain as well as in the United States. Particularly active in this area of research was Bernard Lovell (b. 1913) of the University of Manchester. He did much work on the tracking of meteors and was the first to establish that their orbits are "closed," by which is meant that they are confined to the solar system rather than being of interstellar origin.

Early radio equipment was massive and could not easily be steered to point in a particular direction. Lovell realized the great advantage of receivers that could be conveniently directed, and with great tenacity he organized the construction of a suitable paraboloid radio telescope at Jodrell Bank, Cheshire, near Manchester. The dish was 76 m (250 ft) across and weighed 1,500 tons; it was borne on turrets that had carried fifteen-inch guns on battleships. After many serious financial problems had been overcome, it was finally completed in 1957. The laboratory containing the telescope is now known as the Nuffield Radio Astronomy Laboratory. As soon as it was operating it attracted attention by tracking the carrier rocket of the world's first artificial satellite, the Russian-built Sputnik. The telescope has been used for other satellites as well as for studies of the Moon and planets, but the emphasis has always been on the exploration of outer space by radio astronomy.

Since the establishment of the laboratories at Jodrell Bank, numerous other radio telescopes have been erected in various parts of the world. One of these is the James Clerk Maxwell Telescope, built on top of Mauna Kea, Hawaii, and opened in 1987. Also at Mauna Kea is an optical telescope, the Canada-France-Hawaii Telescope.

Much optical work of great value continues to be done with the 200 in (about 5 m) Hale reflector at Mount Palomar in California. Other notable optical telescopes are the Keck Telescope at Mauna Kea, with an optical diameter of 9.82 m (32 ft), and two 8.1 m (26.6 ft) Gemini telescopes, one in Mauna Kea and the other in Cerro Pachon, Chile. Most of these large telescopes are supported jointly by a number of countries. Perhaps the most impressive of all is the European Southern Observatory's Very Large Telescope (VLT), constructed in Cerro Paranal in the Chilean Andes by a consortium of European nations. It comprises four telescopes each with an 8 m mirror, which together function as a gigantic 16 m telescope.

A significant recent development is the establishment of optical and radio telescopes in space. Valuable pioneering work was done by *Voyager 2*, which made important observations from 1977 to 1989, when it reached the planet Neptune. The Hubble Space Telescope was launched in 1990 but because of a series of technical difficulties only became fully operative in December 1993. It now makes observations in all fields of astronomy. Mention has already been made of results obtained by the satellite *Hipparcos*, operated by the European Space Agency, and primarily concerned with making precise measurements of the positions of stars and galaxies. Its name is partly a tribute to eminent Greek astronomer Hipparchus of Rhodes (c. 170–c. 125 BCE) and an acronym standing for *hi*gh *p*recision *par*allax *co*llecting *s*atellite.

A particularly interesting advance made by radio astronomy was the discovery of *quasars*. The word quasar is an acronym for *qua*si-stell*ar* *r*adio source. Quasars belong to a class of starlike celestial objects that exist at enormous distances from Earth. After their discovery by their emission of radio waves, some of them were also identified by optical astronomy. Their spectra were found to exhibit a large redshift, emissions in the far ultraviolet being displaced into the visible part of the spectrum; this shows that they are moving with enormous speeds. In spite of the fact that in radio emission the individual photons carry very little energy, the energy emitted by the quasars is greater by far than that emitted by any other cosmic source. The energy output is sometimes more than that of 10 billion times that of the Sun, which means that they emit as much energy as an entire galaxy. Matter has become packed such that the nuclei are very close together and the density is enormously high. Quasars are believed to be young versions of galaxies; thus they provide us with an understanding of the very remote past.

Related to the quasars are *black holes,* which are objects consisting of extremely dense matter that create enormous gravitational fields. Anything coming close to a black hole is drawn into it and becomes part of the dense mass, with the evolution of tremendous amounts of energy. The existence of black holes can be inferred from the attraction they exert on other matter. There are good reasons for believing that a black hole with a mass equal to that of millions of stars exists at the center of some galaxies. Our own galaxy appears to have at its center a supermassive black hole, about 2.6 million times more massive than the Sun.

Another discovery of great interest was that of *pulsars*. In 1967 Jocelyn Bell (b. 1943), a research student at Cambridge University working under the direction of Anthony Hewish (b. 1924), noticed radio pulses emitted from a particular point in space at extremely regular intervals of about 1.3 seconds. Later in the year she and Hewish found other objects with similar periodic behavior. Today over five hundred are known. It was later recognized that pulsars are neutron stars, the existence of which had been postulated earlier. Theoretical studies suggest that the magnetic field of an assembly of neutrons the size of Earth is a thousand billion ($10^{12}$) times that of Earth and that the radiation results from this enormous field. The pulsation observed is similar to that of a lighthouse; the star rotates rapidly, and the pulse is observed on Earth corresponding to the passing of the peak of the emission. In 1974 Hewish received a Nobel Prize in physics, sharing the prize with Sir Martin Ryle, whose pioneering work in radio astronomy was mentioned earlier, and it is of interest that this was the first time that a Nobel Prize was awarded for work in astronomy.[5] A feature of pulsars is that although their emission is normally extremely regular, occasionally a "glitch" occurs in which there is a temporary change of frequency. This has been attributed to the fact that a pulsar has a solid crust around a fluid core, and that occasionally there is slippage between the crust and the core.

Radio techniques also made possible the discovery of the cosmic microwave background. Since this particularly relates to understanding the creation of the universe, we will consider it in the next chapter.

The most recent work in astronomy has revealed that the size and complexity of the universe are vastly greater than ever previously imagined. There are 100 billion ($10^{11}$) galaxies, often disk-shaped with a typical diameter of one hundred thousand light-years. To get some idea of a number like $10^{11}$, suppose that each one of them was given a designation consisting of about six letters or numbers. Suppose that we wanted to publish a series of volumes listing just each of the designations, what would the result look like? The answer is that it would consist of something like three hundred sets of all thirty volumes of the *Encyclopaedia Britannica.*

The shapes and sizes of galaxies cover a wide range. There are giant elliptical galaxies, much larger than our own, and some that are much smaller. Our own galaxy has about 200 billion ($2 \times 10^{11}$) stars in it and has a mass equal to the mass of about 400 billion ($4 \times 10^{11}$) Suns; we say that its mass is 400 billion ($4 \times 10^{11}$) solar masses. The total number of stars in the universe is estimated to be about $10^{21}$. It is difficult to comprehend numbers of this magnitude, but the following analogy may help. Suppose that we represented each star by a ball about the size of an orange. If we piled $10^{21}$ of them together, they would occupy a volume of about $10^9$ cubic km (about 240 million cubic mi). If these were piled on the United States, which has a surface area of about 10 million ($10^7$) square km (about 3.8 million square mi), the layer would be about 100 km (62 mi) thick—much higher than any mountain on Earth. So that we realize how insignificant we are (at any rate from a numerical point of view), keep in mind that our Sun is just one of the pile.

Still another way of getting some idea of the size of our universe is to imagine our Sun contracted to the size of an orange. Earth would then be a tiny pellet 20 m (22 yd) away from the orange. On the same scale, the nearest star would be 10,000 km (6,200 mi) away. The galaxy Andromeda, a member of our Local Group of galaxies, would be 5 billion km (3 billion mi) away. The galaxies that are farthest away from us, about 10 billion light-years away, would be 200 million million ($2 \times 10^{14}$) kilometers, or about 20 light-years, away from the orange-Sun! Galaxies contain unusually high concentrations of stars compared with the rest of space, and if every star in the universe were dispersed evenly over the space that has been observed up to now, the nearest star would (on the reduced scale, with the Sun the size of an orange) no longer be 10,000 kilometers away; it would be many millions of kilometers from the orange representing the Sun.

It is estimated that in the universe there are about $10^{78}$ atoms and about $10^{88}$ photons. As we shall see in the next chapter, at the beginning of time there were no atoms but simply a vast number of photons from which the atoms were formed. Today the photons still greatly outnumber the atoms, by a factor of about 10 billion. If all the stars and planets in the universe were split into their constituent atoms, and the atoms were spread uniformly throughout the universe, there would be about one atom in every ten cubic meters of space. The diffuse gas between the galaxies also contains one atom per ten cubic meters. The total density of atoms in the universe is therefore one atom per five cubic meters, or 0.2 atoms per cubic meter. This figure is of great significance, since theory tells us that if the density of atoms in the universe were more than 10 atoms per cubic meter, the gravitational attraction between them would eventually bring the expansion of the universe to a halt. Since the density is well below this critical density, scientists predict that the universe will go on expanding indefinitely.

One odd feature of our universe should also be mentioned. Astronomers have now been able to estimate the amount of matter in the universe, and even its nature. They can also deduce the amount of matter that must be there from the extent of gravitational attraction. It is possible, for example, to deduce the mass of Earth and of the Sun from the speed of planets orbiting around them, and in the same way the mass of galaxies can be deduced from the speed of stars orbiting within them. The problem is that the mass of every galaxy is larger than is accounted for by the material now identified as being present within it. There thus must be some additional, so far undetected, substance present in the universe, without which some of the stars in the galaxies would fly away and the galaxies would fall apart. The additional material that must be present, amounting to some 90 percent of the mass of the universe, is called *dark matter*. Up to now astronomers have no strong opinions as to what it could be. One suggestion has been that the neutrinos, which are known to have a tiny mass, can account for the missing mass. The latest work on neutrinos suggests, however, that they cannot account for more than about a tenth of it. Another suggestion is the missing mass is made up of brown dwarfs, stars that are too small for nuclear reactions to occur in them and that emit so little radiation that they have escaped detection. Some astronomers believe that they have detected one brown dwarf present in our own solar system, moving in an orbit around our Sun but at an enormous distance from it. It may be that there are many brown dwarfs present in the universe, detectable only with great difficulty.

We will discuss more about our Earth and our universe in the next chapter, which is devoted to the big bang, which began everything; the formation of Earth and its age; and some of the processes that are occurring in stars.

8

# OUR CHANGING UNIVERSE

In addition to considering the universe as it is today, we should ask how much we can infer about how it began and how it has developed. All the evidence supports the conclusion that the universe is at least 12 billion years old and that our Earth has existed in its present solid state for about 4.5 billion years. In this chapter we consider the main evidence for these conclusions. It is difficult to visualize these enormous periods of time, and analogies are of help. In his fascinating book *Before the Beginning* (1997), Sir Martin Rees pointed out that if someone were to walk across the United States with a view to arrive at the other side in 12 billion years, he would have to take *one step every two thousand years.* Suppose that some eccentric and lethargic man decides to walk from New York to San Francisco at that pace and that things have been so arranged that each step brings him farther back in time. He thus sees the universe getting younger and younger, as depicted in figure 51. The first single step has taken him back to the beginning of the Common Era. A few more steps, and he is beyond all recorded history. As he travels the first few kilometers, he meets a few creatures that look human, but well before he has gone even one hundred kilometers, there are no more to be seen. As he lumbers on he sees various odd-looking animals and plants, but by the time he is a quarter of the way across the country he sees only some primitive living cells, and after some

**Figure 51.** A trek across the United States would give us some idea of how long ago certain events occurred in our universe. The traveler takes one step every two thousand years, reaching the destination in about 12 billion years. The numbers show, in billions of years, how long ago the events occurred. Note that the time that human beings have existed is negligible compared with the age of the universe.

more steps there are no more. At about one-third of his trek he notices that our Earth is cooling and that rocks are forming. While still in Nevada and not yet in California he sees some stars being formed in our galaxy. For the rest of his trip he is surrounded by a whirling cloud of hot material, and at the end he gratefully collapses into a dense mass at a temperature of many billions of degrees.

Astronomers deal with objects that are very far away, the vast majority of them so distant that we can never make direct contact with them. With the Moon and a few planets it is practicable to bring small samples from their surface back to Earth, but most of the heavenly bodies are completely out of our reach. Information about them can only be obtained by observations from a great distance, such that the science involved is an observational rather than an experimental one. At the same time, theories about astronomy receive much support from laboratory experiments—for example, from work on nuclear reactions. Theories in astronomy are suggested by certain observations, which lead to predictions that can be tested. There is overwhelmingly strong circumstantial evidence for the conclusions about the age of the universe and of the planets. We shall see that the evidence for one particular theory of the creation of the universe, the big bang theory, is very persuasive.

We have discussed some of the contributions of Edwin Powell Hubble, who found that there are many galaxies far outside our own and made a detailed study of these extragalactic systems. His most remarkable observation was that the more distant galaxies showed a larger red shift in their spectral lines than the nearer ones. The relationship that he discovered between rate and distance is a simple one that we call a *proportional* relationship or a *linear* relationship; if a galaxy is twice as far from us as another one, it is receding twice as fast. The speed of recession is thus related to its distance by the formula

$$\text{speed of recession} = \text{Hubble's constant} \times \text{distance}.$$

Because of the enormous distances involved for the remotest galaxies, the value of the Hubble constant is difficult to obtain precisely. Recently, more reliable values have been obtained by observations from artificial satellites.[1]

Galaxies are slowing down as time goes on because of the gravitational attractions between them, but the effect is small and does not greatly affect our estimates of the age of the universe. Suppose that we imagine going back in time, which we can do by imagining that the speeds of the galaxies are all reversed. When we go back in time the galaxies that are far away come back at high speed, the nearer ones at lower speeds, until at zero time they are all in the same place. In other words, at some time in the far distant past, they must have been all in one place. The obvious conclusion is that the universe was created by an enormous explosion in which, as in all explosions, some fragments were propelled much more rapidly than others. The ones that were propelled at the highest speeds have by the present time traveled the longest distances and are at the outer extremities of the universe. The slowest ones have not gone so far. From the size of the Hubble constant it is easy to calculate that the explosion must have occurred at least 12 billion years ago. From now on we will use this conservative figure, even though the true age is probably somewhat greater.

The universe is thus expanding, from highly dense matter that exploded so that the fragments traveled in all directions at immense speeds. The suggestion is not that we are at the center of the universe and that everything else is rushing away from us. We have discussed this apparent expansion of the universe in connection with stretching a piece of rubber; any two marks on the rubber will separate as the rubber is stretched. A rather useful analogy is provided by a lump of dough, with currants in it, that we place in an oven. As it cooks, the whole lump swells, and the distance between every pair of currants increases. In the same way, as our universe expands, the distance between almost every pair of galaxies increases. There are a few exceptions: for example, because of the attraction of our galaxy, the neighboring Andromeda galaxy is approaching us.

One of the first theories of this expansion, referred to at first scornfully but now with respect as the *big bang theory*, was put forward by the Belgian astronomer the Abbé Georges Edouard Lemaître (1894–1966). In 1927, before there was any evidence for an expanding universe, he deduced from Einstein's theory of relativity that the universe might behave in that way. Later, when the idea of an expanding universe had been strongly supported by the observations of Hubble and others, Lemaître developed his ideas in various directions, and it was he in particular who suggested that the universe was originally small and highly compressed. Today this is the most popular theory of the origin of the universe. Some other theories have been proposed, but all but the big bang theory have been rejected by most astronomers.

Several excellent books written about the theory of the big bang go into many details of the physics involved.[2] I think, however, that we can appreciate the theory and be convinced that it must be correct even if we know much less physics than is given in many of these books. Most of what we need has been covered in early chapters of this book. We should remember that all of the chemical elements are conveniently regarded as made up of three basic particles: electrons, protons, and neutrons. Protons and neutrons can be regarded as built from quarks. We should also bear in mind the four types of force that control the behavior of matter; it is convenient to list them again for reference with their approximate relative strengths:

| | |
|---|---|
| • The force of *gravity* | 1 |
| • The *weak nuclear* force | $10^{25}$ |
| • The *electromagnetic* force | $10^{37}$ |
| • The *strong nuclear* force | $10^{39}$ |

In considering the big bang theory, we need to know another bit of physics, relating to temperature effects. Since our universe contains so much energy the big bang must have been accompanied by the production of incredibly high temperatures. Initial temperatures of over a billion billion Kelvin have been estimated, but the exact value does not matter to the argument. The point is that within a tiny fraction of a second there was an enormous expansion, and that afterward the temperature plunged. The important temperature to keep in mind is 3,000 K, because above that temperature electrons could not become associated with atomic nuclei and therefore freely pervaded the universe. Under these conditions, estimated to have lasted for about five hundred thousand years, the universe was opaque, which means that photons (particles of light) could not travel freely because of the electrons in their way. After the temperature had fallen below about 3,000 K, however, the atmosphere became transparent, and light traveled freely. Electrons then became associated with nuclei, and some molecules could be formed by the combination of atoms.

Two lines of evidence, quite different from each other, have provided overwhelmingly strong support for the big bang theory. The first of these, *cosmic microwave background radiation*, was discovered by chance. In the 1960s American astrophysicists Arno Allan Penzias (b. 1933) and Robert Woodrow Wilson (b. 1936) of the Bell Laboratories in Holmdel, New Jersey, were exploring the Milky Way with a radio telescope, their object being to improve communications with satellites. At a wavelength of 7 cm, which is in the microwave region of the spectrum (refer to fig. 29, p. 97), they found a greater effect than could be explained by any known source. They found that the radiation was equally strong from all directions, even coming from apparently empty sky. From the characteristics of the radiation they were able to establish that it was being emitted from a source that had an apparent temperature of about 3 Kelvin, that is, of –270 degrees Celsius. At first they had difficulty understanding the source of this radiation. They went to great lengths to exclude the possibility that it was due to some source on Earth; they removed two pigeons that had tended to deposit their excrement on the antenna, but the pigeons came back (as pigeons do), and they had to be removed more decisively.

The radiation remained, and it took some time to identify its source. It was finally realized, and no better explanation has been suggested, that it is a kind of afterglow, or "fossil," left behind by the big bang. As cooling occurred, the strong forces first came into play and attracted the quarks together, so that atomic nuclei formed. For a time the universe was opaque to radiation, but after about half a million years it had cooled to about 3,000 K. The electrons then became associated with protons and neutrons to form atoms, and the universe became transparent to radiation. The temperature of 3 K for the background radiation tells us that since the time the universe became transparent until the present day it has expanded by a factor of about 1,000, causing the temperature to drop from 3,000 K to 3 K. At the same time, as a result of the stretching of space and in accord with relativity theory, the wavelengths have increased by the same factor. For their discovery of the background radiation Penzias and Wilson shared the Nobel Prize in physics for 1978.

It was realized after their discovery that an earlier observation made by the Canadian astronomer Andrew McKellar could now be understood. In 1941, performing a spectroscopic analysis of outer space, he observed the spectrum of the molecular species CN. It contains one carbon atom and one nitrogen atom, and is hydrogen cyanide, HCN, from which a hydrogen atom has been removed. McKellar found that CN in outer space has a temperature of about 2.4 K. It is obviously another relic of the big bang.

Many further investigations have been made of the background radiation, and much more evidence has been obtained for the big bang. In November 1989 the *Cosmic Background Explorer* (*COBE*) satellite was launched into orbit above the Earth's atmosphere. A microwave recorder measured the intensity of the radiation over a range of wavelengths and in various directions, and the results were found to agree exactly with the predictions of the big bang theory: the average temperature of the radiation can now be given more precisely as 2.736 K. This consistency of the data is so good that there can no longer be any doubt that the background radiation is indeed fossilized radiation left over after the big bang. One refinement of the investigations has been particularly impressive: the big bang theory leads to the conclusion that when the universe had cooled to 3,000 K there were density fluctuations; certain regions, where galaxies were later to form, were of higher density than the rest. Corresponding to these fluctuations there would be, according to the theory, regions of space where the temperatures would be different, but only by one part in about one hundred thousand. These tiny fluctuations over the sky were actually observed in the *COBE* experiments, exactly as predicted by the big bang theory.

The big bang theory also leads to a satisfactory theory of *nucleosynthesis*—the formation of the atomic nuclei—that explains the *distribution of elements* in the universe. We have seen that all atoms are composed according to a similar pattern. Each one of them has a nucleus with electrons orbiting around it, the number of orbiting electrons being equal to the positive charge on the nucleus, and this number defining the identity of the element. The nucleus is different for each chemical element, but it always consists of protons and (except hydrogen) neutrons. There is no reason to doubt that the atoms in a particular chemical element are, and always have been, the same throughout the universe. Our reason for assuming this comes from spectroscopic measurements made on many atoms, both on Earth and in the stars. For example, the spectrum of a carbon atom in a galaxy that is billions of light-years away is exactly the same as that of a carbon atom on Earth. That means that 10 billion years ago, carbon atoms 10 billion light-years away from us were just the same as they now are on our planet, and the same seems to be true of all of the chemical elements.

There are about ninety reasonably stable chemical elements, but six of them play a particularly important role in our universe: hydrogen, helium, oxygen, nitrogen, carbon, and phosphorus. The status of the first two is remarkable, since in the universe as a whole they are by far the most abundant elements. Of all the matter in the universe, roughly three-quarters (by weight) is hydrogen, and one-quarter is helium. (Put differently, about 92 percent of all the atoms in the universe are hydrogen atoms, and about 7.5 percent helium atoms; a helium atom is four times as heavy as a hydrogen atom.) All the rest of the elements together make up only a very tiny proportion, between 1 and 2 percent by weight. Hydrogen is plentiful on Earth but hardly exists at all in its uncombined state. Because of its low density any free hydrogen gas on Earth soon floats to the top of the atmosphere and disappears into space. Helium only exists in the neighborhood of radioactive substances, which produce it as they disintegrate; again, because of the low density of helium, it floats away into the upper atmosphere and beyond. The element helium, which was not discovered until the latter part of the nineteenth century, is highly inert in the sense that it forms no compounds. Hydrogen does readily form chemical compounds that are abundant on Earth, particularly water ($H_2O$). Hydrogen also exists in many compounds that occur in fossil fuels, as well as in all animals and plants.

Carbon has some unique properties that make it particularly significant as far as life in the universe is concerned. Carbon atoms can link together in chains of any length to form molecules of great size and complexity, such as proteins and DNA. As a result, the number of known chemical compounds that contain carbon (*organic* compounds, as they are called) is enormously larger than the total number that do not contain carbon (which we call *inorganic* compounds).

A crucial test of the big bang theory is to see whether it can give a satisfactory explanation for the distribution of elements in the universe. The most striking feature of this is that hydrogen and helium are by far the most abundant and that all the other elements are much less abundant. The explanation is that hydrogen was formed first, and then helium, and that the rest of the elements were formed later from these two elements. The first explanation as to how helium was formed was put forward by Russian-American physicist George Gamow (1904–68) and his colleagues. Gamow was a man with a great sense of humor, and in publishing this work he perpetrated a joke that has become a scientific classic. The research was actually done with his student Ralph Alpher (b. 1921), but Gamow felt that to publish with Alpher alone "seemed unfair to the Greek alphabet." He therefore added to the paper the name of Prof. Hans Bethe (b. 1906), who had already made important contributions to the theory of nuclear reactions in the stars. The paper appeared in 1948 under the authorship of Alpher, Bethe, and Gamow. The fact that the date of the publication was the first of April was a particular delight to Gamow. One detail about the story remains obscure to this day, since both Gamow and Bethe were reticent about it. One story is that Gamow sent Bethe a draft of the paper asking if he would agree to his name being added and that Bethe had no objection. The other is that Bethe did not know that his name would be added and that at first he was not at all pleased but later took the matter in good part.

Born in Odessa, Gamow was educated at the University of Leningrad, where he later became a professor of physics. After research at both Göttingen and Cambridge he moved to the United States, and from 1934 to 1955 he was professor of physics at George Washington University in Washington, DC. After that he spent his last years at the University of Colorado. He created the character of Mr. Tompkins and in a number of popular books used him to communicate to the general public an understanding of relativity, particle physics, and quantum mechanics.

The essence of the theory is as follows. First, protons became associated with neutrons, to form deuterium (refer to fig. 5, p. 38), which they could do readily at the enormous temperatures existing as a result of the big bang. Deuterium atoms, however, are uncommon in our universe, constituting only about one in every ten thousand hydrogen atoms, and are thus much less common than helium atoms. The reason is that the high temperatures of the early universe would have caused them to unite rapidly to form helium. We have already discussed this fusion reaction, in which two deuterium nuclei come together to form a helium nucleus (two protons and two neutrons):

$$^{2}_{1}H + ^{2}_{1}H \rightarrow ^{4}_{2}He + 23.7\ MeV.$$

At the high temperatures existing shortly after the big bang this process occurred readily, its driving force being the large amount of energy evolved in this fusion process. As a result, practically every deuterium nucleus formed immediately after the big bang was converted into a helium nucleus. It can be deduced from the big bang theory and our understanding of nuclear physics just why the amount of hydrogen in the universe is almost exactly three times the amount of helium. This alone is a very convincing argument in favor of the theory.

The remaining chemical elements were not produced in the earliest stages of the development of the universe because of the exceedingly high temperatures, at which these elements are unstable. They were instead made in the stars, where the temperatures are lower but still high enough to induce fusion reactions. The formation of carbon is of special interest not only in view of its great importance in living systems but also because for a time it presented a scientific problem that was finally resolved by British cosmologist Sir Fred Hoyle (1915–2001). A carbon nucleus has six protons and six neutrons, so it could be formed from three helium nuclei, each of which has two protons and two neutrons:

$$^{4}_{2}He + ^{4}_{2}He + ^{4}_{2}He \rightarrow ^{12}_{6}C.$$

However, three helium nuclei are unlikely to collide together at one time, even in the dense atmosphere of a star. It is much more likely that two helium nuclei would first come together to form a nucleus containing four protons and four neutrons, which is a nucleus of the element beryllium:

$$^{4}_{2}He + ^{4}_{2}He \rightarrow ^{8}_{4}Be.$$

There seemed to be a serious difficulty, however. The $^{8}_{4}Be$ nucleus has never been detected, even in unstable form; the form $^{7}_{4}Be$, containing only three neutrons, is the stable form of the element. It therefore seemed that there was little chance that the $^{8}_{4}Be$ nucleus would exist long enough to allow it to combine with another deuterium atom to form carbon:

$$^{8}_{4}Be + ^{4}_{2}He \rightarrow ^{12}_{6}C.$$

Hoyle realized, however, that there would be a way out of the difficulty if the carbon nucleus happened to have a particular characteristic. If the $^{8}_{4}Be$ and $^{4}_{2}He$ nuclei coming together in the reaction just written happened to have an energy corresponding to an energy level in the carbon nucleus, their combination to form the carbon atom would be much more likely, and might take place before the unstable $^{8}_{4}Be$ atom had time to disintegrate. This speeding up of processes as a result of a coincidence of energies is referred to by physicists as *resonance*. When Hoyle had his idea, the energy levels of the carbon atom had not been measured, and he arranged for the necessary experiments to be done; a suitable energy level was indeed found. The formation of carbon in this way, in the stars, is therefore quite plausible. It is interesting that the proof of this possibility, which strongly favors the big bang theory, should have been obtained by Hoyle, who had always been a strong critic of the theory, preferring a "steady-state" theory of the universe, in which it is eternal, having no beginning and no end. As the universe expands, according to this theory, new matter is created to keep the average density the same. The new microwave evidence and the interpretation of the distribution of the chemical elements, however, have left that theory with hardly any supporters.

The heavier elements were made in the stars in similar ways, by fusion processes. The matter was gone into in great detail in a paper titled "Synthesis of Elements in Stars," which was published in 1957 in the *Reviews of Modern Physics*. The authors were Margaret Burbidge

(b. 1922), her husband Geoffrey Burbidge (b. 1925), William Alfred Fowler (1911–95) and Fred Hoyle; the paper is now known to astronomers as "$B^2FH$." It remains one of the classic papers of science, still fundamentally sound after so many years. One point is particularly worthy of mention. The oxygen nucleus $^{16}_{8}O$ is formed by the fusion of another helium nucleus to the carbon nucleus:

$$^{4}_{2}He + ^{12}_{6}C \rightarrow ^{16}_{8}O.$$

If the oxygen nucleus also had a matching resonance energy, this reaction would have proceeded so fast that little carbon-12 would have remained. If that had been the case, life in the universe would never have occurred. The existence of amounts of carbon and oxygen that have made living things possible is thus due to a delicate balance of forces: there was a suitable resonance energy for the carbon nucleus, but not for the oxygen nucleus.

One interesting feature of nucleosynthesis is of special interest and significance. It turns out that an isotope of iron, iron-56 or $^{56}_{26}Fe$, is energetically more stable than both elements whose atoms that are lighter than the iron atom and those that are heavier. In other words, in an energy diagram of the elements, iron is at a minimum. As a result, iron is formed more easily, and therefore in larger amounts, than some of the lighter elements and all of the heavier ones. The fact that the abundance of iron in the universe is relatively high gives strong support to the $B^2FH$ theory of how the elements were formed.

The stars, including our Sun, are nuclear fission reactors in which a vast number of nuclear reactions occur, providing them with their energy. Most of the elements produced in the stars remain there, but occasionally a star explodes at the end of its life cycle (an event called a *supernova*) and expels elements. Aside from this, there is a steady flow of material thrown out by stars; we know this from studies of the solar wind issuing from our Sun.

▲▲▲

We have not yet considered how the stars and our Earth were formed. After the universe was first created by the big bang, an enormous mass of gas expanded at high speed, accompanied by powerful shock waves. At first, little was present except hydrogen and helium gases, the latter formed by the nuclear processes just mentioned. From time to time a massive cloud of gas suddenly shrank as a result of gravitational attraction, forming a dense blob of spinning matter at a very high temperature. The oldest galaxies and stars, which contain the lowest proportions of the heavier elements, were formed in this way. In the course of time, as a result of matter being thrown out by stars exploding at the end of their life cycles (in supernovae) the gas gradually became more enriched with heavy elements.

Our galaxy was probably created about 10 billion years ago. Initially it was just gas and dust swirling in a complicated fashion and becoming disk-shaped. Stars, some with their attendant planets, were formed in it from time to time. About 5 billion years ago our Sun and its surrounding solar system condensed from swirling gas and dust. The lighter material surrounding our Sun floated to the top, and eventually some of it condensed into the planet Jupiter, which is composed of gas at high pressure; it probably has a solid core of rock and ice. The planets Saturn, Neptune, and Uranus were probably formed in a similar way. These four planets, all of which have disk-shaped systems of rings, are the largest of the outer planets of our solar system.

The heavier elements in the Sun became concentrated in the inner regions of our solar system and condensed into the inner planets, one of which is Earth. Our planet, like the other planets, was therefore created from discarded material from stars going through their life cycles between about 10 billion and 5 billion years ago. As a result of the manner of its formation, Earth is significantly enriched in the heavier elements, which are necessary for the creation of rocks. In addition, it began with enough carbon, as well as hydrogen in the form of water and other compounds, for life to become possible.

▲▲▲

Earth was at first molten and existed in that state for many millions of years before it became a solid mass somewhat resembling what it is today. Evidence from rocks on Earth leads to conclusions about its age and tells us how it came to have its present form.

For many years the age of Earth was a matter of some controversy. We now know its age to be more than 4½ billion years, but until the end of the nineteenth century many investigators assumed ages that are much shorter. Until about 1750, in fact, most of them thought that the Old Testament version was essentially correct: Earth was considered to be less than six thousand years old, all sedimentary rocks being deposited during the Great Flood, in which Noah played the leading role. Other surface features of Earth were assumed to have resulted from catastrophic events like floods that occurred from time to time.

As geological observations were made, however, many of the investigators realized that such short periods were quite inconsistent with the evidence. Many considered that a few hundred million years were necessary rather than just a few thousand. One of the first persuasive proponents of the longer periods was Scottish geologist James Hutton, whose work, carried out in the late eighteenth century, we noted earlier. He became convinced that changes to Earth's surface came about gradually over enormous periods of time. He emphasized the erosive role of rivers in excavating valleys and in depositing dissolved and suspended material into the sea. Sediment carried along by rivers would be washed into the sea and would accumulate in deposits, which might form new rocks by the action of heat from within the planet. He could not avoid the conclusion that many millions of years, perhaps some hundreds of millions, would be needed for mountains to be formed and be eroded.

Charles Darwin, who did much geological as well as biological work during his famous voyage on HMS *Beagle* in the 1830s, also estimated that geological events were taking place hundreds of millions of years ago. He was particularly interested in a wide valley known as the Weald of Kent in southeast England, near his family's later home. Darwin believed that this valley had arisen from the encroachment by the sea on the line of chalk cliffs in the south of Kent. He estimated that the encroachment would occur at the rate of one inch per century, and from the configuration of the valley he concluded that it had been formed 300 million years ago.[3] Darwin, of course, realized that the planet itself may have been formed very much earlier.

At about the same time, distinguished geologist Sir Charles Lyell, whom we met earlier, also made estimates from the same kind of evidence as well as from the speed with which sedimentary rocks might have been formed. His estimates, and those of many other contemporary geologists, were also in the hundreds of millions of years, in agreement with Darwin.

These estimates, although we now know them to be much too low by a factor of at least 10, were strongly criticized by those who believed that the estimates from the scriptures were to be taken literally. There was also strong dissent from an unexpected quarter, namely from William Thomson (later Lord Kelvin), whom we have met in connection with the laws of thermodynamics. Between 1862 and 1899 he wrote a series of papers in which he calculated the time it would take for Earth to cool to its present temperature from its temperature when first formed in a molten state. He took into careful consideration the luminosity of the Sun, the expected rate of cooling of Earth, the formation of Earth's solid crust, and even the effect of lunar tides on the rate of Earth's rotation. His initial conclusion, put forward in 1862, was that Earth could not have solidified more than 100 million years ago. In later publications he proposed even shorter times, a few tens of millions of years. Kelvin called particular attention to Darwin's estimate of 300 million years, which he thought was impossibly high, since he concluded that Earth would not even have been solid at that time; the presence of surface water, and the formation of valleys, he thus considered impossible.

Kelvin's estimate, of course, is still much more than suggested in religious writings but far less than many geologists then believed to be correct. Kelvin thus succeeded in displeasing everyone. His opinion was disturbing in view of his great reputation, particularly because he was recognized as an authority on heat. Some geologists gave much consideration to whether the geological events could have taken place much faster than previously thought. Some of them thought that there might be another source of energy in the planet that kept it warm for much longer periods, and Kelvin himself had pointed out this possibility as early as 1862, referring to "sources of heat—now unknown to us . . . in the great storehouse of creation." It seems likely that Kelvin was being sarcastic in making this comment, believing the idea to be highly improbable.

This then unknown source of heat turned out to be the solution to the dilemma. In 1897 radioactivity was discovered. It soon became apparent that there was a vast store of radioactive substances in the planet. Radioactive disintegrations are accompanied by emission of heat, which helped to keep Earth warm for a longer period. Kelvin's upper limit of 100 million years is thus far too low. From the amount of radioactive material in the planet, estimates were made of their heating effect, which was found sufficient to explain how Earth could be several billion years old.

The study of radioactive substances later also contributed more directly to making reliable estimates of the age of rocks and other materials in the planet. One important radioactive process is the disintegration of uranium-238, which is the isotope present in largest proportion (about 99.3 percent) in natural uranium:

$$^{238}_{92}\text{U} \rightarrow {}^{4}_{2}\text{He (an } \alpha\text{-particle)} + {}^{234}_{90}\text{Th}.$$

(Remember what this equation tells us. The nucleus $^{238}_{92}\text{U}$ has 92 protons and 146 neutrons, or 238 nucleons. If it loses a helium nucleus, $^{4}_{2}\text{He}$, it loses 2 protons and 2 neutrons and becomes $^{234}_{90}\text{Th}$, which has 90 protons and 144 neutrons, i.e., 234 nucleons.) The half-life of the process is about 4.5 billion years, which makes it particularly convenient for measuring the ages of rocks that are a few billion years old.

Estimates of the age of a rock can thus be based on the fact that as uranium disintegrates it produces helium gas, which is chemically inert and stable. Being a gas, helium easily leaks out of molten rock, but after the rock cools and solidifies this gas remained trapped. An analysis of a sample of rock for the amount of uranium and the amount of helium trapped will therefore, by a simple calculation, tell us how much time has elapsed since the rock solidified. Experiments of this kind have been done many times, and the final result is the estimate that Earth has existed in its present, solid state for about 4.5 billion years. Some rocks were indeed formed at later times, but their age has never been found to be less than several hundred million years. The estimates of the geologists were therefore shown, by entirely independent measurements, to have been by no means too high.

This particular technique, of determining the amount of helium present in uranium minerals, was used by Ernest Rutherford. He obtained ages of about 500 million years for the particular rock samples he used and thus proved conclusively that Lord Kelvin was wrong. In 1904 Rutherford presented a lecture at the Royal Institution on the age of Earth and on the production of heat by radioactive substances. When he entered the lecture room he was somewhat disconcerted by the fact that Kelvin, then aged eighty, was in the audience. Rutherford reported later that Kelvin at first fell asleep but that when he began to speak about the age of Earth, he saw "the old bird sit up, open an eye and cock a baleful glance" at him. With great presence of mind Rutherford modified the thrust of his lecture somewhat, emphasizing that in 1862 Kelvin had referred to "sources of heat now unknown to us" and that he had qualified his estimate of the age of Earth by writing, "provided no new source of heat is discovered." "Thus," said Rutherford with admirable tact, "the audience must admire the foresight, almost amounting to prophesy, which had made Lord Kelvin so qualify his calculations." Rutherford was then relieved to see that "the old boy beamed" on him.

Besides Rutherford, British geologist Arthur Holmes (who also did valuable work on continental drift) was a great pioneer in determining geological time scales. In 1913, when he was only twenty-three years old, he published a book titled *The Age of the Earth*, which soon established itself as a scientific classic. It contained the passage, "It is perhaps a little indelicate to ask our Mother Earth her age, but Science acknowledges no shame and from time to time has boldly attempted to wrest from her a secret which is proverbially well guarded." The book is a masterly account of the results of radioactive work and of how they related to the more classical studies on rates of erosion and sedimentation. Since that time work on radioactivity has been greatly extended, without requiring any significant changes to Holmes's early conclusions.

There are other ways in which uranium disintegration can be used to determine the ages of rocks. The $^{234}_{90}Th$ nucleus, which is a product of the disintegration, is itself radioactive, and the reaction cited above is only the first step in a chain of radioactive processes, all of which occur more rapidly than the original step. The final product of the chain of processes is lead-206, and the whole chain of processes can be written as

$$^{238}_{92}U \rightarrow\ ^{206}_{82}Pb + \text{several particles.}$$

Since the first step, with a half-life of 4.5 billion years, is the slowest in the chain, it controls the overall rate, so we can take the half-life of the whole chain to be 4.5 billion years. American chemist Bertram Borden Boltwood (1870–1927), who worked for a period with Rutherford and was for some years professor at Yale, greatly extended techniques of this kind for the dating of rocks.

Suppose that, while a planet or a meteor was still a molten mass thrown out from the Sun, some uranium was trapped in molten rock. At first the uranium could move fairly freely in the liquid rock, but after the rock solidified it would be trapped, and the lead formed would be forced to remain near the uranium. Thus, from the ratio of the amounts of uranium-238 and lead-206 found in any part of the rock, we can calculate the time that has elapsed since the rock solidified. Various types of meteorites have been found on Earth, and this technique has been applied to a number of samples. The meteorites classified as "stony" are the oldest, having an age of about 4.6 billion years. These are the oldest known objects in the solar system, so that at least some of the solar system must have solidified 4.6 billion years ago. Rocks brought back to Earth from the Moon by the Apollo astronauts have also been found to be of a similar age.

Many other radioactive substances have been used to determine the ages of rocks. Only one other example need be given here: this is the rubidium-strontium method. Rubidium, chemical symbol Rb, is a fairly rare metal similar in its properties to the much more common metals sodium and potassium, both of which are widely distributed in minerals. Rubidium forms no minerals of its own, but it is so similar to potassium that it can substitute for potassium in all potassium-containing minerals. It occurs in easily detectable amounts in minerals of this type, such as the micas and certain clay minerals.

Rubidium has two naturally occurring isotopes, ${}^{85}_{37}Rb$ and ${}^{87}_{37}Rb$, their relative abundances being 72.2 percent and 27.8 percent, respectively. The latter is radioactive, emitting a beta particle to become an isotope of strontium, ${}^{87}_{38}Sr$. The element strontium has properties similar to the much more common element calcium and can replace it in many minerals. Strontium has four naturally occurring isotopes, the particular isotope ${}^{87}_{38}Sr$ normally occurring to the extent of 7.04 percent. However, when strontium occurs in the neighborhood of rubidium-containing rocks, its percentage is higher. Measurements of the relative amounts of strontium ${}^{87}_{35}Rb$ and ${}^{87}_{37}Sr$ in such circumstances lead to estimates of the age at which the rocks solidified. The results are always consistent with those obtained in other ways, giving ages of several billion years for the oldest rocks, with shorter ages for some metamorphic rocks, which underwent a fiery transformation at a later time.

Many other examples could be given. There are several radioactive elements that allow ages to be determined and many cases where isotope ratios in rocks provide such evidence. The science of isotope geology is now highly developed, leading to a coherent picture of the ages of various geological events. As a result, the circumstantial evidence from isotope geology indicating that Earth is at least 4.5 billion years old is overwhelmingly strong. When it is added to all the other evidence leading to the same conclusion, it is hard to see how anyone could dispute it.

▲▲▲

The method by which energy is emitted by the stars, including our Sun, was also a puzzle until nuclear processes were well understood. More, however, is involved than the radioactive disintegrations that help to keep Earth warm. Since high temperatures exist in the stars, fission and fusion processes also play essential roles. Before considering these matters we should look briefly at some of the procedures that are used for obtaining information about stars. All of the stars except the Sun are visible to us as mere points of light, and for most of them the light takes many millions of years to reach us. Nevertheless, astronomers have been able to discover a great deal about the inner workings of stars; several techniques are used to gain information about them. Reliable estimates can be made of their distances from us, and we now have a good idea of the age of many of them. The mass of the Sun can be determined precisely from the way its gravity influences the orbits of its several planets. The masses of stars that have companions—those that form binary systems—can also be determined, less precisely, from the way they move around one another.

Spectroscopic measurements on stars provide a variety of information: chemical composition, surface temperatures, energy emitted, and the speeds with which the stars are receding from us. Since every chemical element has its own unique set of spectroscopic fingerprints, the elements present in the stars can be identified without question. Their amounts can also be estimated. It is now known, for example, that our Sun is composed of about 70 percent (by weight) hydrogen, 28 percent helium, and only about 2 percent of all the heavier elements. This is quite similar to the composition of the universe at large, which is something like 75 percent hydrogen, 25 percent helium, and less than 1 percent of all the other elements.

There is an interesting story behind the present knowledge that the composition of the Sun and stars is similar to that of the universe at large and quite different from that of Earth. For a time it was believed that Earth had split off from the Sun, and if this were the case the composition of the Sun would be much the same as that of Earth. The work of Cecilia Payne-Gaposchkin (1900–79), however, gave the first indication that this was not correct. Born Cecilia Helena Payne in England, she showed great aptitude for science and mathematics at an early age. In 1919 she went to Newnham College, Cambridge, where she attended lectures by Arthur Eddington (more on him below), some of them on his confirmation of relativity theory; these lectures particularly inspired her to become an astronomer. After graduating from Cambridge she was awarded a fellowship to Radcliffe College to do research at the Harvard College Observatory, where she was to spend her entire career. For her doctorate at Harvard she studied the spectra of stars, including the Sun, and was able to deduce for the first time reliable pressures, temperatures, and compositions for a variety of stars. The important conclusion at which she arrived was that all the stars have remarkably similar compositions, consisting predominantly of hydrogen and helium, with only small amounts of the heavier elements. Since it was then firmly believed that Earth and the Sun must have similar compositions, her conclusion was hotly disputed. Her research director at Harvard, renowned astronomer Harlow Shapley (1885–1972), was particularly skeptical of her conclusion, and one of her examiners, the equally famous Henry Norris Russell (1877–1957), insisted that she add a significant qualification to her thesis before he would accept it. She was required to write, with regard to hydrogen and helium, that "the enormous abundance derived for these elements in the stellar atmosphere is almost certainly not real." It later became clear that it was certainly real. Cecilia Payne-Gaposchkin obtained her PhD in 1925 and distinguished herself in her later career; she did much more research in astronomy and in 1956 became the first woman professor at Harvard.[4]

Many elements were discovered for the first time from their spectra, and helium provides a particularly interesting example. On August 18, 1868, a total eclipse of the Sun was visible in India, so a number of scientists had gone there to make observations. One who examined photographs of the spectra was Joseph Norman Lockyer (1836–1920), who, though a civil servant at the British War Office, had already done valuable work in his spare time in astronomical spectroscopy. He was particularly interested in a so-called $D_3$ line in the yellow region of solar spectra. It was known that the well-known sodium D line was in fact two lines close together, called the $D_1$ and $D_2$ lines. Since the $D_3$ line could not be obtained from any substance available in the laboratory, Lockyer boldly suggested that it was caused by a new element, found in the Sun but apparently not on

Earth. He gave this new element the name helium, from a Greek word meaning "the Sun."

Over a quarter of a century later, William Ramsay (1852–1916), who with Lord Rayleigh had already discovered the element argon, began in 1895 to investigate the gas produced by a radioactive mineral called cleveite. On examining its spectrum he found a line that he remembered, from a lecture by Lockyer that he had attended many years previously, to have been called the $D_3$ line and to have been identified as relating to a new element. He sent a sample to Lockyer, who was then director of the Solar Physics Laboratory in South Kensington, who then turned over his whole laboratory to the study of cleveite. The existence of helium was confirmed, and soon afterward, in 1897, Lockyer was knighted by Queen Victoria; Ramsay was knighted in 1902 and received the Nobel Prize for chemistry in 1904 for his discovery of inert gases and his investigations of their chemical and physical properties.

Lockyer continued to make significant contributions to spectroscopy, but perhaps his most memorable achievement was his founding in 1869 of the journal *Nature,* which he edited for the first fifty years of its existence and which continues to play a distinguished role in the communication of science.

Spectra play other important roles in science. Through the Doppler effect they allow us to determine accurately the speed with which stars are moving away from us, which nearly all of them are doing. We saw earlier, in connection with the distribution of the energies of molecules, that temperature influences the form of the distribution curve. As a result, the spectrum of a star allows the temperature of its surface to be estimated.

During the nineteenth century, scientists began to think about the physics of stars and about their ages. Particular attention was paid to the Sun, for that is the star we know most about. Geologists concluded that Earth had existed for at least many hundreds of millions of years, so the Sun must be a good deal older. This, however, presented a problem, because it was obvious that no ordinary chemical reaction could keep the Sun hot for anything like that length of time. If the Sun were a solid lump of coal and was burning in pure oxygen vigorously enough to generate the amount of heat that the Sun does, it would not have survived for more than about fifteen hundred years.

Lord Kelvin and Hermann von Helmholtz, both of whom we met particularly in connection with the laws of thermodynamics, tackled this problem. Their suggestion, which turned out to be only part of the answer, was that gravity plays an important role. If the Sun started out as a thin cloud of gas, the force of gravity would make it more and more dense and compact. Its potential energy would therefore decrease, and according to the first law of thermodynamics there would be evolution of heat. When Kelvin and Helmholtz made the calculations they obtained ages of 100 million and 25 million years, respectively. Since Kelvin had estimated that the age of Earth was somewhat less than 100 million years, he thought that his estimates were satisfactorily consistent. However, as we have seen, the geological evidence strongly favors much greater ages for the rocks on Earth.

When radioactivity was discovered it was first thought that radioactive disintegrations contributed to the heating of the Sun. In 1903 English astronomer William Watson calculated that if there were 3.6 grams of pure radium in each cubic meter of the Sun's volume, the heat evolved in the radioactive decay would be enough to supply all the heat radiated by the Sun. Similar ideas were developed by astronomer George Darwin, one of the sons of Charles Darwin. After the work of Cecilia Payne-Gaposchkin had been confirmed, however, scientists realized that the Sun could not contain enough radioactive material for this explanation to be acceptable.

Instead, other types of nuclear processes provided an explanation of the long lives of the Sun and other stars. Even before Cecilia Payne-Gaposchkin had done her work, Arthur Stanley Eddington (1882–1944; see fig. 52) had essentially given the correct explanation. He was born in 1882 in Kendall, Westmoreland, and first took a degree from Owens College, Manchester, which is now part of the University of Manchester. From there he went to Trinity College, Cambridge, where he graduated in 1904 as Senior Wrangler (top of the class) in the Mathematical Tripos. He was elected a fellow of

**Figure 52.** Sir Arthur Eddington (1882–1944), British astronomer and cosmologist, is generally regarded as the founder of the science of astrophysics. He was the first to formulate a satisfactory theory of the physics of the stars, a theory that took into account the nuclear reactions occurring in them.

Trinity but soon left Cambridge to take the position of chief assistant at the Royal Observatory in Greenwich. In 1913, at the age of only thirty-one, he was appointed Plumian Professor of Astronomy and Natural Philosophy (i.e., physics) at Cambridge. In the following year he also became director of the University Observatory.

In about 1916 Eddington became particularly interested in the physics of the stars. Four years later he gave a lecture that is considered to have constituted the birth of the science of astrophysics. As we have seen, Eddington had at the time become a great expert on the theory of relativity and appreciated the implications of Einstein's famous formula $E = mc^2$. In 1920 no fission or fusion process had yet been discovered, but Eddington had already realized that they were possible and that under the conditions present in the stars they might well occur and contribute to extending the lives of the stars.

His idea was that a special role was played by the conversion of hydrogen nuclei into helium, although at the time the details of how this could happen were still not understood. It was known that the universe consists largely of hydrogen with about a third as much helium, and it was thus natural to conclude that there must be some mechanism by which helium atoms were being produced from hydrogen atoms. From Einstein's equation Eddington was able to deduce that when four hydrogen nuclei come together to form a helium nucleus, much energy is liberated, the mass balance (using modern values) being as follows:

| | | |
|---|---|---|
| Mass of four protons | $= 4 \times 1.00794$ | $= 4.0318$ u. |
| Mass of one helium nucleus | | $= 4.0026$ u. |
| Loss of mass | | $= 0.0292$ u. |

(It should be recalled from chapter 5 that the unit *u* is the *atomic mass unit,* defined in such a way that the mass of the isotope $^{12}C$ [carbon-12] is exactly 12 u.) Einstein's formula $E = mc^2$ leads to the result that 1 u corresponds to an energy of 931.5 MeV (million electron volts). It follows that a decrease in mass of 0.0292 u will lead to the production of 27.20 MeV of energy. (Remember that the *electron volt,* eV, is the energy acquired by an electron when it passes through a voltage drop of 1 volt.)

On this basis, and from the amount of energy that the Sun is actually radiating, Eddington was able to estimate that the brightness of the Sun can be maintained if about four million tons of the Sun's matter is being converted into energy every second, by the formation of helium atoms from hydrogen atoms.

That is roughly $10^{15}$ tons of matter per year; that seems a great deal, but it is only a tiny fraction of the mass of the Sun, which is about $2 \times 10^{27}$ tons. To convert that amount of matter into energy, the mass of hydrogen that has to be converted is about 600 million tons of hydrogen per second, or about $2 \times 10^{16}$ tons of hydrogen per year. In 10 billion ($10^{10}$) years it will have consumed about $2 \times 10^{26}$ tons, or 10 percent of its entire mass. This conclusion is quite consistent with the estimate that the Sun has an age of about 5 billion years and that it will probably survive in its present form more or less for another 5 billion years.

As we have seen, Kelvin and Helmholtz pointed out that gravitational energy would be converted into heat. Eddington's calculations, however, showed that such a process could keep the Sun hot only for a few tens of millions of years, not for the billions of years required by the geologists (4.5 billion years as we now know). In 1920 he expressed very clearly the idea that nuclear processes must be playing a dominant role. By so doing he deduced the occurrence of these nuclear processes before any of the fission or fusion reactions had been discovered. He even made a prediction about nuclear energy, saying that one day we might be able to release it and put it to practical use. Within twenty-five years this had been achieved.

Eddington pursued this line of investigation, showing that from the basic laws of physics one could make important deductions about the nature of stars. If a star is at equilibrium, neither expanding nor contracting, the inward pull of gravity must be balanced by the outward pressure. This is partly the pressure resulting from the radiation emitted by the hot gases and partly the pressure that results from the kinetic energy of the motions of the atoms and molecules present. Knowing the mass of a star and with certain other information, we can calculate from the laws of physics what its size and temperature must be and how much energy it radiates. For many stars it is thus possible to be able to deduce an energy balance. In particular, we can infer the amount of energy that is radiated from the surface of the star. This is something that has been measured directly, and Eddington's predictions were found to be largely correct.

At the temperatures that exist inside stars, electrons are stripped off atoms, resulting in a so-called *plasma* of positively charged particles. The fast-moving particles emit radiation that exerts pressure, called *radiation pressure*. If a ball of gas has more than a certain critical amount of mass, the fast-moving particles generate so much radiation pressure that the gas is blown apart. Eddington worked out a theory of this and concluded that there are three possible fates for a hot ball of gas. If it is too small, a plasma cannot form, and it will become a cool mass in which there is no radiation pressure, so the inward gravitational attraction is balanced by only the gas pressure; the planet Saturn, for example, is a body of this kind. If it were somewhat bigger, it could become a stable glowing star like our Sun, with the gas pressure plus the radiation pressure balancing the gravitational pressure. If it is bigger still, it will be unstable, shining brilliantly before being blown apart by the enormous radiation pressure. By reasoning in this way, Eddington was able to work out the range of masses in which stars can exist. Even though today we know much more about the details of what stars are made of, and of what nuclear processes can occur, modern estimates of the range of masses are much the same as Eddington's.

Eddington also reached the conclusion that the centers of all stars, whatever their mass and brightness, must have much the same temperatures, now estimated to be 15 to 20 million degrees. The reason that all stars have a similar internal temperature is *negative feedback*. This is what is involved in the thermostats that control the temperatures of our homes. If the temperature is too high the thermostat turns the furnace off, and the house cools down; when it gets too cold the furnace goes on again. There is a similar effect in a star: if it shrinks a little, the increased gravitational attraction will cause a release of energy, causing the star to get hotter. That in turn causes the star to expand until equilibrium is restored. If, on the other hand, the star expands, the reduction of gravitational attraction will produce cooling, which causes it to contract again. A star thus has a built-in thermostat that keeps its core at a fairly constant temperature. This temperature is high enough for nuclear reactions to occur, enabling the star to live much longer than it would otherwise.

In 1926 Eddington summed up these basic conclusions in an important book, *The Internal Constitution of the Stars*. For quite a few years Eddington's ideas were dismissed by many physicists, but later it was realized that he had been essentially right all along. It may seem surprising today that Eddington never received a Nobel Prize for his remarkable work; at the time, however, astronomy was not regarded by the Nobel officials as a branch of physics: the first Nobel Prize in physics for work in astronomy was not given until 1974, when awards were made to Hewish and Ryle.

▲▲▲

We should now consider in more detail the role played by nuclear reactions in the stars. At first sight we would think that a nuclear reaction occurring at the heart of a star would increase its temperature; however, because of feedback, this is not what happens. Without nuclear reactions, which produce energy and therefore provide an outward pressure to resist the inward pull of gravity, the star would shrink and therefore get hotter still. The nuclear reactions thus have the effect of preventing the collapse of the star as a result of gravity and therefore prolong its life.

One of the objections that physicists raised to Eddington's ideas about the interior of stars was that the nuclear reactions could not occur sufficiently rapidly. One of the simplest of the nuclear reactions is the fusion of two deuterium nuclei to give a helium nucleus. Deuterons, however, repel each other strongly, and Eddington's critics objected that even at the enormous temperatures in the stars, deuterons would not have enough energy for them to come together and undergo this process; much higher temperatures would be required for them to do so. Eddington pointed out that helium does exist to a considerable extent in the universe. In his own words, "The helium that we handle must have been put together at some time and some place." He then added, perhaps tongue in cheek, a comment that is often quoted: "We do not argue with the critic who urges that the stars are not hot enough for this purpose; we tell him to go and find a hotter place." This is usually interpreted as Eddington's polite way of telling his critics to "go to hell."

Soon after Eddington's book appeared, the answer to the apparent dilemma was suggested by George Gamow (of Alpher, Bethe, and Gamow, discussed earlier), who made many important contributions to the understanding of nuclear processes in the stars. In 1928, just two years after Eddington's book appeared, Gamow applied the new quantum mechanics to nuclear processes. Heisenberg had proposed the principle of uncertainty in 1926, and the new quantum mechanics allowed the possibility of what is called *quantum-mechanical tunneling*. The idea behind this is as follows:

suppose that a vehicle needs to surmount a hill and that a certain amount of energy will allow it to do so. If the vehicle approaches the hill with too little energy it cannot surmount it. This is certainly true for a vehicle to which classical mechanics applies, but for particles on the atomic scale the situation is different. Quantum mechanics does allow a particle of atomic size to get from one side of a hill to the other even if it does not have enough energy to surmount it; we say that the particle *tunnels* through the hill.

This is the point that Gamow made in his 1938 publication. Two deuterons can interact with one another, admittedly with a low probability, even though they do not have enough energy to surmount the energy barrier. This explanation successfully avoided the difficulty that appeared to make the nuclear processes impossible. Gamow also showed in the same investigation that in spite of the strong repulsive force, two protons can readily join together and eliminate a beta particle (an electron) to form a deuterium atom. Since a proton can easily combine with a neutron, it follows that wherever there is ${}^{1}_{1}H$ in a star, there is always a supply of deuterium, ${}^{2}_{1}H$. At the high temperatures of stars these fuse together so rapidly to form helium that the proportion of deuterium to ordinary hydrogen remains low.

▲▲▲

Let us finally look briefly at the life cycle of a typical star. Much of what we conclude about stars does not come to us directly from observational evidence but from computer simulations of what occurs in stars, based on the knowledge of the laws of physics. Support for these deductions comes from actual experiments carried out on Earth—for example, to determine the way in which certain nuclear reactions occur under the conditions believed to exist in stars.

As a result of investigations of this kind, a whole body of knowledge about stars has now been constructed. Everything that has been deduced forms such a coherent package that it is difficult to believe that it can be far from the truth. The kind of information we can obtain about a star is in brief as follows: if we know the mass of a star and the temperature at its surface, the basic laws of physics along with computer models allow us to calculate how hot the star is at its interior and what its pressure and density must be. Then, knowing the temperature at the center of the star and its composition, we can conclude something about the nuclear reactions that must be going on in the interior. Also, from experiments carried out on Earth, we know how much energy is being released in the star. This can then be compared with the amount of energy that the star actually does radiate. In this way we can confirm that our model of processes taking place in the star is self-consistent.

Star formation is going on all the time, by the recycling of material from clouds of gas and dust in space. When a star is formed it is surrounded by clouds of material, some of it in the form of spinning discs, from which planets may form. Star formation occurs when such a cloud gets squeezed from outside, perhaps by shock waves. It then starts to collapse as a result of gravitational forces, and as it collapses it fragments, forming binary stars and other structures. The initial pressure that causes the collapse is due to compression waves that move around the galaxies, perhaps as a result of an exploding star. A compression wave, of which a sound wave is an example, squeezes the gas as it passes but leaves individual molecules more or less undisturbed after it has passed. The largest stars that are formed in this way have only relatively short lifespans (no more than a few million years, rather than about 10 billion for our Sun).

A star like the Sun, however, has a size that is conducive to longevity. After it has formed and has shrunk by gravitational attraction until its temperature has reached about 15 million degrees, it begins to operate as a nuclear fusion reactor, converting hydrogen into deuterium and then into helium. As long as hydrogen remains in the core, these nuclear reactions keep the star from reacting further and getting any hotter. The lifetime of such a star, which is known as a *main-sequence star*, depends on its mass. The bigger it is, the less time it spends as a main-sequence star, because it burns its fuel more vigorously to keep from contracting and getting hotter. A star twenty-five times more massive than the Sun spends only about 3 million years as a main-sequence star. We have seen that by contrast the Sun will last a total of about 10 billion years and that since it has already shone for about 5 billion years, it is about halfway through its life as a main-sequence star. A star half as massive as the Sun may last about 200 billion years.

When all of the hydrogen in the core of a star like the Sun has been converted into helium, meaning that its nuclear energy supply is depleted, its core will begin to shrink and generate heat as gravitational energy is released. The rise in temperature will make the outer part of the star expand and turn the star into what is called a *red giant*. When this happens some matter is blown away into space and the star swells up, in the case of the Sun, to almost the size of the orbit of Venus. During this time the temperature of the core will reach a temperature of about 100 million degrees, and at such temperatures helium will be converted into carbon. It will then become stabilized as a red giant, which will survive for about a billion years, after which its helium supply is exhausted.

At the end of this phase the carbon core begins to collapse, releasing enough heat to allow hydrogen to burn for a further period, farther out from the center of the star. During this period the star will expand to a size roughly equal to the present orbit of Earth, at the same time releasing elements such as nitrogen and carbon into space. Nuclear processes are now unimportant because of lack of fuel, and eventually the inner core, now largely carbon, simply cools down. The star eventually settles down as a white dwarf and will be about the size of Earth. Its density will therefore be enormous, since today the volume of the Sun is roughly a million times that of Earth. The density of the white dwarf will thus be about a million times that of water; a teaspoonful of it would have a mass of over a ton. Consistent with this conclusion, no white dwarf has been observed that has more than 1.4 times the mass of the Sun. This fits in with the inference from theory that a star that has such a mass has a significantly different life cycle. A heavy star will go through more stages of nuclear burning, at higher and higher temperatures, and will manufacture more of the heavier elements such as carbon and iron, as well as still heavier elements such as uranium. What happens in the case of a star that is more than 1.4 times as heavy as the

Sun is that the gravitational compression is so great that many of the electrons present combine with protons to form neutrons, by a process that is the reverse of the beta decay of a neutron:

$$^{1}_{1}\mathrm{H} + {}^{0}_{-1}\mathrm{e} \text{ (an electron)} \rightarrow {}^{1}_{0}\mathrm{n}.$$

In this way the star is reduced to an enormous mass of neutrons, which, since they do not repel one another, can be packed together closely. It thus forms what is called a *neutron star*, which is much more dense than a white dwarf. We have seen that a white dwarf with the mass of the Sun has the size of Earth; a neutron star with 1.5 times the mass of the Sun would be much smaller, with a diameter of only about 10 km (6.2 mi). A teaspoonful of it would weigh about a billion tons. There is a limit to the size of a neutron star, since if the mass is too great the gravitational forces are large enough to cause it to shrink indefinitely and form a *black hole*. The gravitational field in a black hole is so strong that nothing can escape from it, not even light.

In the early chapters of this book, we saw how our understanding of the universe was advanced by the scientific study of atoms, molecules, and energy forms on Earth. In the present chapter, we have seen how the study of stars, far away and of massive size, has helped our understanding of the origin of the universe. We are now ready to discuss the life forms in it.

9

# LIVING THINGS

One very important feature of the universe is that life exists in it. Living systems are distinguished from inanimate ones by their capacity for continual growth, development, and reproduction. At one time it was believed that living organisms were imbued with a special "vital force," by virtue of which they were able to perform their functions. It was often supposed that living systems, and especially human beings, were not bound by the same laws of nature that apply to inanimate objects. For example, it was often supposed that the evolution of species, in which systems experience an increase in order, is a violation of the second law of thermodynamics.

It is not possible to prove that such ideas are wrong, but the understanding of living systems that has come from recent scientific work has made them unlikely to be true. Nothing that we know about biological systems is incapable of explanation in terms of the same laws of nature that have been found to apply in all of the branches of science, such as physics and astronomy, that apply to inanimate systems. We thus have no reason to question the beauty and harmony of scientific understanding.

▲▲▲

The science of biology, which deals with living things, is now advancing at an enormous speed. Until the twentieth century it was still mainly concerned with the collecting of information, and although there were some well-established theories, the subject lacked the precision of physics and chemistry. This is due to the very nature of the subject: biological systems are remarkable in showing many variations from one to another simply because of their great complexity compared to most of the systems studied in any of the other sciences. Since biological information is so extensive we must confine ourselves here to a few topics. I will focus attention on the evidence relating to the evolution of species and discuss how the theory of natural selection developed. Then we will see how since the middle of the twentieth century the subject has become transformed by the introduction of physical methods, such as x-ray determinations of the structures of some large and important biological molecules.

One of the first biologists was Aristotle (384–332 BCE), whom Darwin much admired. Aristotle collected and recorded a vast body of biological information, particularly relating to animals. He did a certain amount of classification (taxonomy) but placed no emphasis on the subject and nowhere listed the main taxonomic groups he had recognized. He would not have favored the idea of the evolution of species.

Over the centuries a great deal of work was done on the classification of plants and animals. Living things were divided into numerous groups, the members of each group having marked similarities to each other. Each group can be divided into subgroups, and there is further subdivision into species. By definition, the members of a species can interbreed, whereas members of different species cannot. By today's reckoning there are about 30 million known species of plants and animals on Earth. Their distribution is quite remarkable: there are about 10,000 different species of birds and 20,000 species of ants. More remarkably still, there are at least 300,000 species of beetles. Distinguished British geneticist J. B. S. Haldane was once asked what the study of biology tells us about God. Haldane replied, "He must be inordinately fond of beetles." Since there are so many species of ants, and so many ants on Earth that they comprise one-half of the mass of all insects, God must also be very fond of ants.

At first the general assumption, suggested by religious writings, was that every species on Earth was created separately and independently. The various and numerous species of animals and plants were considered to be immutable, having been placed on Earth in their present form. A different point of view was suggested by French naturalist Georges-Louis Leclerc, Comte de Buffon (1707–88). From 1749 to the end of his life, he published twelve volumes of his work, *Histoire Naturelle*. As he collected his information he was struck by the fact that there are marked similarities between members of different species, for example between the dog, wolf, fox, and jackal. This made it likely that they were members of one family and most unlikely that they had been created separately, especially in view of anatomical features that seemed to serve no useful purpose. Here, obviously, lay the germ of a theory of evolution. A more explicit theory of evolution was put forward in the 1790s by Erasmus Darwin (1731–1802), a man of great insight, who was also Charles Darwin's grandfather.

Another theory of evolution was proposed by Frenchman Jean-Baptiste-Pierre-Antoine de Monet, Chevalier de Lamarck (1744–1829), whose name is now commonly associated with a major error. This is unfortunate: his work exerted an important influence on others, including Darwin. Most of his work was done at the Natural History Museum in Paris. He arranged animals in a linear series, beginning with primitive organisms and continuing to the most complex. Lamarck's suggestion in about 1809 was that there is a gradual process of development from the simple to the complex; he went wrong, however, in suggesting *how* the changes come about. He believed that changes brought about in the lifetime of an individual organism would be inherited by the next generation. In other words, he proposed the *inheritance of acquired characteristics*, which is now usually known as *Lamarckism*. His theory is simply explained by the example of why giraffes have long necks. In order to survive, they eat the leaves of tall trees, and therefore stretch their necks. The next generation would, according to Lamarckism, inherit the longer necks, and from generation to generation the necks would get longer.

Lamarck's career was an unhappy one: he was first a soldier, then a bank clerk, later an assistant botanist, and finally a professor of insects and worms in Paris. He suffered from poverty all his life and went completely blind. Poverty pursued him even after death, since his remains were put into a rented grave, and five years later, when the tenure expired, his body was disinterred and his remains dispersed. It is regrettable that famous French naturalist Georges Cuvier (1769–1832) ridiculed Lamarck in a sarcastic eulogy, in which he described his ideas as worthless and unscientific. In fact, Lamarck's work was of good quality, and he was the founder of the field of invertebrate biology, a term he invented. His conclusion about the inheritance of acquired characteristics was not unreasonable, being accepted by Darwin at first. We now know that under some special circumstances, acquired characteristics can be passed on to future generations. For example, if a person is exposed to x-rays, mutations often occur and are transmitted to the next generation. Also, it is now known that the genes of a developing embryo can be affected by chemicals, such as drugs, in the surrounding fluid, and the next generation may be affected. Lamarck's idea was in reality a perfectly sensible one; the inheritance of acquired characteristics can occur under some special conditions but is not the main factor in the evolution of species.

The theory that is now generally accepted was proposed later in the century by Charles Robert Darwin (1809–82; see fig. 53), whose conclusions were based on his observations during a five-year voyage as a naturalist on a British naval vessel, HMS *Beagle*. On this voyage Darwin gained a detailed practical knowledge of the geology and biology of many distant lands. He made a vast collection of geological and biological specimens, which he shipped back to England. On his return home in 1836 he thought deeply about his results for many years and published many of his observations. However, he was reluctant to publish his ideas about evolution, as he disliked controversy and knew that they would produce something of a sensation. At first he had accepted Lamarck's idea of the inheritance of acquired characteristics but later decided that an important change was required. Darwin's alternative explanation for the long necks of giraffes was that if at a given time some giraffes had longer necks than others, the ones with the longer necks were more likely to survive,

**Figure 53.** Charles Robert Darwin (1809–82) ranks as one of the greatest scientists of all time. His theory of evolution by natural selection has had a profound influence on the progress of biology and on the lives of everyone today—including those who claim not to accept the theory.

since they could reach more food and were more likely to have offspring. Those with shorter necks could reach less food and were less likely to survive long enough to produce offspring.

Darwin's decision to publish his theory of evolution was occasioned by his receipt in 1858 of a letter from Alfred Russel Wallace (1823–1913), a younger man who had also made a lengthy sea voyage. The letter revealed to Darwin that Wallace had come to much the same conclusions as he had about the way in which evolution occurs. Wallace was obviously about to publish his theory, which would have won him priority. In the end Darwin and Wallace came to an amicable agreement, which was a great credit to both of them. Papers written by Darwin and by Wallace were read at a meeting of the Linnean Society in London on July 1, 1858. Oddly, neither of the two men was present on this historic occasion, and even more oddly the announcement appears to have been largely ignored by other scientists.

The following year Darwin published his master work, *The Origin of Species by Means of Natural Selection.* In 1871 appeared his *Descent of Man and Selection in Relation to Sex.* His theory at once attracted much attention and created considerable controversy among the public, although it was fairly quickly accepted by many scientists, who understood the strength of the evidence for it. Today it is Darwin rather than Wallace who is chiefly remembered for the theory; Darwin's exposition of his theory was indeed more complete and was supported by more evidence than that of Wallace.

Darwin's theory has often been referred to as the *survival of the fittest,* this expression having been invented by Herbert Spencer (1820–1903). Spencer was a man of remarkable versatility, who had been a railway engineer, teacher, and journalist before devoting himself to writing on science, psychology, and philosophy. At first Darwin did not like the expression, preferring the term *natural selection,* which was used in the title of his first book. His change of opinion was prompted by the realization that, as Spencer had pointed out, natural selection was being misinterpreted as meaning that God controlled the process, the exact opposite of what Darwin meant. Darwin in fact did not believe in any such control, although he was reticent to express his views, since he knew that they would produce more controversy.

Darwin's theory is now largely accepted by scientists, since the evidence for his general idea is overwhelming; there is disagreement only about minor details. Much more remained to be done after the theory was put forward: in particular, the theory provided no mechanism by which evolution occurs. Significant light was shed on that question by experiments carried out from 1856 to 1863 by Gregor Mendel (1822–84; see fig. 54),[1] in the gardens of a monastery in Brünn (now Brno), then the capital of Moravia and now in the Czech Republic. Between 1856 and 1863 he grew about thirty thousand garden pea plants, artificially fertilizing certain plants that had special characteristics.

For example, he cross-bred green peas with yellow peas and found that all the peas of the first generation (referred to as $F_1$) were yellow. However, when the $F_1$ peas were selfed (i.e., fertilized with

**Figure 54.** Austrian priest and botanist Gregor Mendel (1822–84), who discovered the basic statistical laws of heredity and provided a mechanism for Darwin's theory of evolution. (Edgar Fahs Smith Collection, University of Pennsylvania Library)

other $F_1$ peas), three-quarters of the peas in the next generation, $F_2$, were yellow, and one-quarter was green. Similar numerical ratios were found when he crossed and selfed plants that showed other characteristics, for example when he worked with sets of tall and dwarf pea plants.

Mendel formulated a hypothesis to explain the vast number of results he had obtained, particularly to explain the ratios of simple numbers, such as ¾ and ¼, that he had obtained. His main conclusion was that each plant receives one factor from each of its parents, and he called these factors "elements," later to be called *gametes*. Often one of these factors was dominant, and another recessive. In the case of the yellow and green peas, for example, yellow was dominant and green recessive. With the tall and dwarf pea plants, tallness was dominant and dwarfness recessive. (Mendel's hypothesis is explained in more detail in figure 55.)

For many years Mendel's work remained unnoticed by other biologists; Darwin had in his library a copy of his publication, but

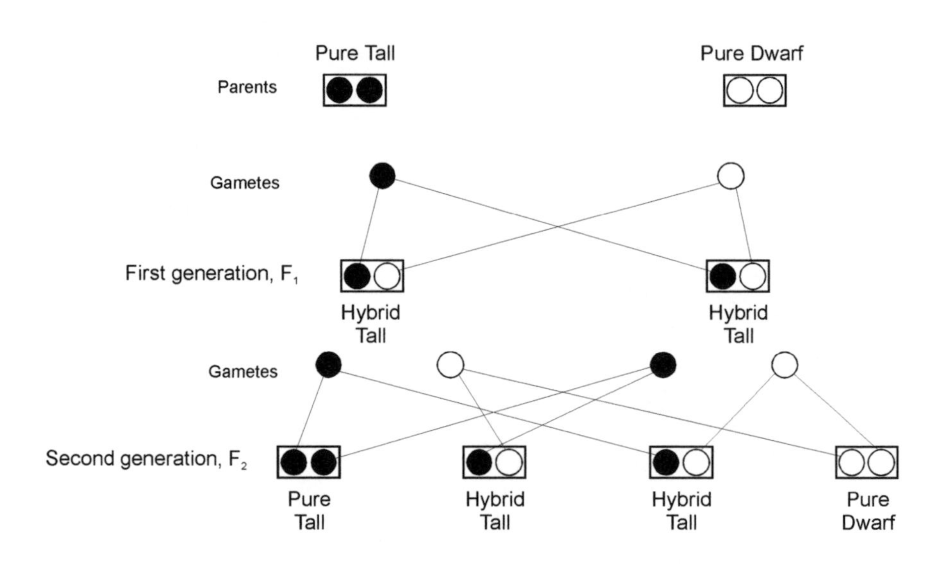


**Figure 55.** Mendel's hypothesis as it explains his experiments on the breeding of tall and dwarf pea plants. He proposed that the difference between the pure tall plants and the pure dwarf plants was due to their containing different "elements" (now called gametes), occurring in pairs. The solid circles in the pure tall plants denote the tall gametes, which he concluded to be dominant; the open circles in the pure dwarf plants are the recessive gametes. When the pure tall plants are crossed with the pure dwarf plants, one tall gamete pairs with a dwarf gamete, and every resulting first generation ($F_1$) hybrid is sure to have a tall-dwarf pair. Since tallness is dominant, the plant is tall. In the formation of the second generation ($F_2$) the pairings can appear in four possible ways, which are shown in the figure. Three of these lead to a tall plant, one of which is a pure tall, the other two hybrid talls (because of the tall dominance). The fourth is a pure dwarf. Of all the plants in generation $F_2$, therefore, three-quarters are tall and one-quarter dwarf, which is just what Mendel found. Similar explanations apply to many other situations, such as the breeding of green and yellow plants as well as the inheritance of hemophilia in humans.

since the pages remained uncut he obviously had not bothered to read it. In more recent years Mendel has been accused of cheating, because his published results fit the expected ratios too precisely. It may simply be that one of his assistants was too anxious to give him the results he wanted. The important point is that whatever the explanation, Mendel was right in his conclusions, as has been shown by a great deal of independent work. The laws of inheritance, however, are often much more complex than Mendel observed.

Much other evidence helped to build up an understanding of how biological inheritance takes place. In 1858, two years before Darwin's *Origin of Species* was published, German pathologist Rudolf Virchow (1821–1902), working at the University of Berlin, showed that each living cell has a central concentration of material known as a *nucleus*. Soon afterward various investigators found that every cell nucleus contains structures called *chromosomes*, which always come in pairs. The pairs double before cell division and are then shared between daughter cells. The obvious conclusion, which later turned out to be correct, was that chromosomes are the carriers of hereditary factors.

Of particular importance were experiments performed from about 1908 onward by American biologist Thomas Hunt Morgan (1866–1945). He carried out, first at Columbia University and then at Caltech, an extensive series of investigations on *Drosophila*, tiny fruit flies that have the advantage for scientists of breeding rapidly, a new generation being produced about every two weeks. Morgan confirmed and extended many of Mendel's findings and observed that certain pairs of genes tend to be inherited together more frequently than other pairs. He concluded that genes are individual units arranged in a particular order along a chromosome, like beads on a string. He was thus able to achieve a certain amount of mapping of genes along a chromosome. Morgan was awarded the 1933 Nobel Prize for physiology or medicine.

In the nineteenth century scientists realized that genes sometimes undergo a change, known as a *mutation*. American geneticist Hermann Joseph Müller (1890–1967) worked among other things on the action of x-rays on cells, showing that they induce genetic mutations. For this work he was awarded the 1946 Nobel Prize for medicine. It was he in particular who called the public's attention to the dangers of various kinds of high-energy radiation. Previously, x-rays had been used much too freely in hospitals, without proper precautions, and x-ray equipment was even used in shoe stores to test whether a shoe fit a customer.

A very important development came in 1944 from the Rockefeller Institute in New York. There Canadian-born scientist Oswald Theodore Avery (1877–1955) and some of his colleagues were working on *pneumococci,* in particular on the transformation of a nonvirulent form into a virulent form. They found that the transformation is performed by a class of substances called deoxyribonucleic acid (DNA). This led to the suspicion that genes are portions of DNA molecules, which occur in long chains, and this was subsequently found to be true of the genes of every living organism, whether plant or animal. James Watson and Francis Crick later determined the structure of a typical DNA molecule; this became of central importance in molecular biology.

At about the same time, significant work was being done on *bacteriophages,* viruses that infect bacterial cells. One scientist who worked in this field was German-born biophysicist Max Delbrück (1906–81), who had been educated as a nuclear physicist and who had studied in Copenhagen with Niels Bohr; much of his work was performed at Caltech. Another scientist who worked on bacteriophages was Alfred Day Hershey (b. 1908) of the Carnegie Institution in Washington. At Indiana University, Italian-born scientist Salvator Edward Luria (1912–91) obtained high-quality electron micrographs of bacteriophages. Delbrück, Hershey, and Luria all independently obtained results that supported the conclusion that genes are portions of DNA molecules, for which they shared the 1969 Nobel Prize for physiology or medicine.

▲▲▲

It was clear by the 1950s that the very long DNA molecules play a crucial role in the evolutionary process. Genes had been shown to be portions of DNA molecules, and it was known that a single gene carries much genetic information, in the form of the order in which four chemical groups, known as *bases,* are arranged along the gene's molecular chain. It was obviously of the greatest importance to determine the structure of a DNA molecule and especially to find out how the four bases are arranged in a particular molecule. By this time physicists were perfecting the method of x-ray spectroscopy to determine the structures of large biological molecules. For example, Dorothy Crowfoot Hodgkin (1910–94) had determined the structure of penicillin during World War II. Various scientists had determined the structures of protein molecules, and in 1948 American chemist Linus Pauling (1901–94) made the particularly fruitful suggestion that helical (spiral) structures sometimes occur in protein molecules.

The story of how the structure of DNA was arrived at is a remarkable one, quite different from the stories of how the structures of other large molecules were determined. In almost all other cases, much painstaking experimental x-ray work was carried out, followed by detailed calculations that today are made much easier by computers. Usually the structure was finally discovered at the end of a careful experimental investigation that provided precise physical information. With DNA, on the other hand, the solution was reached making less use of the x-ray data. Instead, a plausible structure was suggested partly on the basis of the experimental results, but with much consideration given to how the various atomic groups fit together. In particular, the building of models played a vital role.

Four people made significant contributions to this research. X-ray work of the highest quality was carried out by Rosalind Elsie Franklin (1920–58), who did her experiments at King's College, London. Also doing excellent x-ray work on the problem at King's College was Maurice Hugh Frederick Wilkins (b. 1916), who was born in New Zealand. At the Cavendish laboratory at Cambridge was Francis Harry Compton Crick (b. 1916). All of these three had received most of their education and experience in physics. The fourth person was American James Dewey Watson (b. 1928), who was primarily a biologist working at the Cavendish laboratory. The director of that laboratory was Sir William Lawrence Bragg (1890–1971), who had shared the 1915 Nobel Prize for physics with his father for their pioneering work on x-ray crystallography. At first Bragg did not encourage the speculations of Crick and Watson on the structure of DNA, especially since Crick, who had not yet obtained his doctorate, was neglecting the work he was supposed to be doing. Later, however, when the DNA work promised to lead to a structure, Bragg changed his mind.

Crick and Watson had been impressed by Pauling's conclusion that proteins can have a helical structure, and they guessed that DNA would have the same kind of structure. Some of Rosalind Franklin's excellent x-ray photographs, which regrettably had been seen by the others without her knowledge, were consistent with a helix. They were also aware of some of the DNA work done by biochemists. In particular, Erwin Chargaff (b. 1905), at Columbia University, had shown that there was a definite relationship between the amounts of the four bases that occur in DNA. The bases are adenine (A), guanine (G), thymine (T), and cytosine (C), and it seemed that the number of units of A was always equal to the number of units of T, while the number of units of G was the same as the number of units of C. The ratio of the numbers of A and G bases (and obviously the ratio of the numbers of T and C bases), could however vary widely.

Crick and Watson spent much time building molecular models, just as Pauling had done to arrive at his helical structure for proteins. They paid particular attention to the sizes of the various groups, to make sure that everything fit together satisfactorily. In 1953 they proposed that the DNA molecule involves a double helix, with two helices intertwined, and with the base units A, T, G, and C inside the helix. The four bases are of different sizes, and for the double helix to be of uniform width, they concluded that an A unit can only associate with a T, and a G must pair with a C. This pairing was consistent with Chargaff's rule that the number of units of A was equal to the number of T units, whereas the number of G units was the same as the number of C units. They arrived at the pairing because only in that way would the helix be uniform in width; any other arrangement would give bulges along the chain, which was excluded by the x-ray work. Another salient feature of their double helix, introduced mainly to explain how replication can occur, is that the chains run in opposite directions.

Watson and Crick suggested that replication (copying) involves the unwinding of the helix, with each single helix serving as a template for the formation of another helix. This is represented in figure 56. Wherever there was an A in the old strand, a T would be formed in the new strand, and vice versa; a G would always form a C, and a

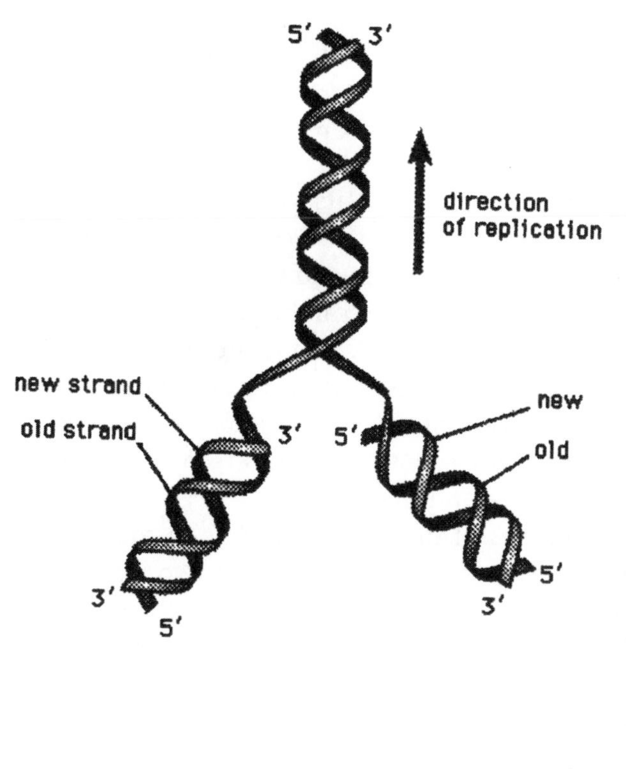


**Figure 56.** A schematic illustration of replication. The double helix of the old DNA strand continuously unwinds into two helices, and new single helices are formed, each complementary to the helix on which it is growing. Helix 1 forms a new helix 2, and helix 2 forms a new helix 1. For chemical reasons the ends of the chains are designated 3' and 5', the two chains running in opposite directions. The two new strands finally pair up, and the resulting DNA molecule is identical to the original one.

C would always form a G. Helix 1 from the original double helix would thus bring about the formation of a new helix 2, while the original helix 2 would form a new helix 1. The two strands of DNA would mirror each other, and replication would be complete.

The announcement of the double helix structure aroused great interest and excitement. The work, in fact, has been called the most original of the twentieth century. Although the experimental evidence was slender, the model seemed to many scientists to be so plausible that it could hardly be wrong. There were some skeptics, however, and alternative structures were proposed over the next few years. However, as the experimental evidence accumulated, the Watson-Crick model became universally accepted by biologists. Crick, Watson, and Wilkins shared a Nobel Prize in 1962; Rosalind Franklin, whose experimental work had been so crucial, had died of cancer four years earlier. Only much later did she receive appropriate recognition; her Cambridge College, Newnham, for example, has dedicated a residence in her name, and in the year 2000 King's College, London, attached her name, together with that of Wilkins, to a new molecular biology building.

The helical structure proposed by Watson and Crick for DNA soon led to rapid advances in the field of *molecular genetics*, which is the application of the methods of physics and chemistry to genetics. Only a bare outline of molecular genetics can be given here. We have seen that chromosomes, which occur in animal and plant cells and transmit hereditary information, consist of strings of genes, which are segments of the long DNA molecules. A gene may be defined as a specific section of a DNA molecule that forms a recognizable unit that plays a role in inheritance and metabolism.

Of all of the known types of molecules, DNA is in many ways the most remarkable. They perform not one, but two important functions, both of them crucial to life. In the first place, they are concerned with the *replication* (copying) of an organism. Secondly, they control the organism's day-to-day survival. Once an organism is born its cells proliferate, and in each one of them is a copy of the individual's set of genes, known as the *genome*. There is nothing intrinsic to a specific gene, in the sense that outside the cell there is no distinction between a bacterial gene, a plant gene, and a human gene; we would not be able to say, after analyzing an isolated gene, which species it came from. Variants of particular genes are known as *alleles* (alternative forms of the same gene). Differences in alleles account for the small amount of genetic variation among different members of the human race (other than gender differences).

We now know that the "working" molecules of a living system are not the chromosomes but the protein molecules. A protein molecule is a long-chain molecule in which there is a chain of amino acid molecules, of which there are twenty types. Each individual protein has a unique sequence of amino acids. Many biological structures, such as muscle and skin, are made up primarily of proteins. It is mainly the proteins that are concerned with physiological activity. In the human body, for example, the protein hemoglobin transports oxygen from the lungs to the places where it is needed. Enzymes, which are proteins, are the biological catalysts. Most of the chemical processes that occur in living systems would occur much too slowly if the appropriate enzyme were not present.

The problem that molecular geneticists had to attack was how the information contained in the genes is able to control the manufacture of proteins. The mystery of how this is done began to be solved in the 1960s. The order in which the bases occur in DNA is the critical feature. Suppose, for example, that in a particular DNA molecule, the order of the four bases T, C, A, and G happens to be TTCGGTCGC etc. One possibility might be to consider the bases in pairs; to indicate this we can put commas between the pairs: TT, CG, GT, CG, etc. This, however, does not work; there are twenty different amino acids, and if we take four bases in pairs we can get only $4 \times 4 = 16$ possibilities:

TT, TC, TA, TG
CT, CC, CA, CG
AT, AC, AA, AG
GT, GC, TA, TG.

Suppose that instead we consider triplets, so that with commas the sequence becomes TTC, GGT, CGC, etc. With triplets we can get $4 \times 4 \times 4 = 64$ possibilities, more than enough. This turned out to be the right answer; often different triplets lead to the same amino acid.

Important contributions were made by South African–born Sydney Brenner (b. 1927) who joined the staff of the Medical Research Council Laboratories in Cambridge in 1957 and for a time worked closely with Crick; he shared the 2002 Nobel Prize for physiology or medicine. Brenner introduced the word *codon* to refer to a triplet of bases in DNA. The essential feature of the coding is that there is an exact correspondence between the sequence of the codons in the DNA of the gene and the sequence of amino acids in the protein that is produced. For example, the codon TTC specifies the amino acid called phenylalanine, GGT specifies glycine, and CGC specifies arginine. The sequence shown above would therefore lead to the synthesis of a protein that would begin with phenylalanine-glycine-arginine. All twenty amino acids could be specified (or coded) without using all of the combinations, but in fact several different codons lead to the same amino acid. Glycine, the simplest of the amino acids, is for example given by any of the following four codons: GGT, GGC, GGA, and GGG. The function of certain codons is to make the synthesis start or stop.

Several people were responsible for breaking the genetic code—that is, in relating the codons to particular amino acids. We have already noted Crick's and Brenner's major contributions; Severo Ochoa (b. 1905) of New York University and Arthur Kornberg (b. 1918) of Stanford University discovered the mechanisms of synthesis of DNA and ribonucleic acid (RNA), for which they were awarded the 1959 Nobel Prize for physiology or medicine; Marshall Warren Nirenberg (b. 1927), of the National Institutes of Health in Bethesda, Maryland, was the first to identify a codon with a particular amino acid; Har Gobind Khorana (b. 1922) of the University of Wisconsin and Robert William Holley (1922–94) of Cornell University also took some of the final steps in unraveling the genetic code, and the three of them shared the 1968 Nobel Prize for physiology or medicine.

Mention should be made of an interesting project begun in 1963 and successfully continued for many years. It was suggested by Sydney Brenner that it would be useful to focus attention on a tiny nematode (roundworm) composed of only 959 cells. This work became known as the Worm Project, and by intensive investigation it was possible by 1986 to work out the main genetic and physiological details of the organism.

The outline I have given has necessarily left out a great many details. The DNA in a cell nucleus does not form proteins directly but stays where it is and makes a copy of itself. In this way the information is transmitted through an intermediate called messenger-RNA (mRNA). Ribonucleic acid (RNA) molecules, of which mRNA is a special form, are also vital constituents of cells. They are formed from DNA in such a way that the bases are arranged in an order equivalent to the order in which they appear in the DNA molecule, but some details of their structure are different.

One additional and essential detail of protein synthesis should be emphasized. X-ray studies of protein molecules have always shown that a given protein normally exists in a particular three-dimensional form; the one-dimensional strands are always folded in the same way. In the globular proteins the long molecules are folded up like a ball of wool, which raises the question (a troubling one for the early workers) of how a molecular template could directly produce such a complicated three-dimensional structure, in a precisely specified pattern. The answer is that it does not do so directly. The protein molecule is sometimes synthesized as a long strand, and when it is surrounded by water molecules, the forces are such that the molecular strand folds itself spontaneously into a particular shape. It is the sequence of amino acids that determines the folding that must take place in the protein. Thus, the biological synthesis of a protein creates the molecule as a long strand of amino acids. Sometimes, as it is being formed, it steadily folds into the unique three-dimensional shape that is a necessary consequence of the particular sequence of amino acids; in other cases it folds into its final shape at a later stage.

Each protein has a specific conformation when it folds itself naturally, and if a protein in water is deliberately unfolded it will rapidly spring back into its specific folded conformation. We would expect to be able to reproduce these folding processes on a computer in order to understand just what forces bring about the folding. However, this does not appear to have been achieved yet for any protein, even at the cost of many hours of computer time. Obviously a delicate balance of forces is involved in the folding processes, and we still do not understand the details. The same must be true of the folding of chromosomes in order for them to fit into cells.

The amount of information that can be stored in a single cell is stupendous. One cell in a human body, for example, can store about ten times as much information as is contained in all thirty volumes of the *Encyclopaedia Britannica*. A bacterial cell has a much smaller capacity, by a factor of about 1,000, and could only store the information contained in the New Testament. Richard Dawkins has suggested a prolific way in which the New Testament could be replicated. It could be encoded into a bacterium, which would reproduce and form 10 million copies a day. He points out a few serious snags, however: encoding the bacterium might take many years, and bacteria are extremely hard to read![2]

Of particular interest to us are the genetic structures in a human cell. Its nucleus contains a string of DNA units distributed over 23 pairs of chromosomes. The length of DNA from one of these chromosomes is about 5 cm (2 in). This means that the entire length of the 46 chromosomes, if laid end to end, is about 2 m (6 ft). The neat packing that is required to get the 23 pairs of chromosomes into the nucleus is best appreciated by means of an analogy. Imagine magnifying the nucleus to the size of an average suitcase, which means that we must magnify it roughly a million times. A single chromosome would then have a length of about 50 km (30 mi) and a thickness of about a millimeter (0.04 in). Imagine packing a suitcase with (among other things) a few pieces of string a millimeter thick and 50 km long! That is what the cell does, and it does it in such a way that each one of the codons in the 46 filaments is accessible and able to carry out its functions in the cell.

Biologists are fond of pointing out that if all the DNA in all the cells in a single human being were stretched out, it would reach to the Moon and back 8,000 times (or sometimes they say to the Sun and back 250 times). Even more remarkable in the human cell is that the 46 chromosomes can be packed into the tiny nucleus and that when packed they can still carry out so much chemistry.

All human beings of the same sex, regardless of race, have the same genes, of which there appear to be about forty thousand. The Human Genome Project, centered in North America and supported internationally, explores some of the details of human genetics. Each human genome contains about 3 billion base units.[3] It may be helpful to consider the enormity of this number in the following way: suppose that we imagine magnifying the genome so that it is 100 km (about 62 mi) long. In order to accommodate the large number of bases, each centimeter of its length would contain about 300 of them, or each inch would contain about 750 of them.

Only a small proportion of the DNA chains are functional, in the sense of encoding protein production. Most of the base pairs—perhaps 98 percent of them—seem to be junk, in that they do not code for anything and perform no useful function at all, being simply errors of copying or undiscarded garbage from the evolutionary process. These junk base pairs are called *introns* by the experts, in contrast to the *exons*, which perform coding. Thus, if we were to travel along the 100 km (62 mi) road that represents the genome we would find the trip quite tedious, in that only a total of 2 km (1.25 mi) would be of any interest to us. A stretch of about 30 kilometers of the road consists of sequences that are simply repetitions of previous sequences, and another stretch, of about 5 kilometers, consists of countless repeats of blocks of five or more units. Although these introns appear to have no functional significance, they are of value in identifying people by genetic fingerprinting, since they cause each individual to be unique.

The first objective of the Human Genome Project was to identify all of the 3 billion base pairs. This part of the project was completed early in the year 2000, so now we have a genetic map of a human being in which every one of the three billion individual base pairs is identified. If this were to be published in the format of all the thirty volumes of the *Encyclopaedia Britannica* it would be about ten times as long. To most of us this map would appear to be dull reading, but some experts find it of absorbing value and interest. It is a list of base pairs belonging to an average person; individuals will, of course, show small variations in base pair composition. In the meantime a number of genes have been identified. The gene that causes about half of us to be males, for example, starts with the following codons: GAT, AGA, GTG, AAG, CGA, etc. There are 240 letters in the complete gene, which is situated on the Y chromosome. Humans who lack this Y chromosome, and therefore this male gene, are females.

Some scientists have been critical of the Human Genome Project, particularly because it has proved very expensive and has siphoned off money that might have been used for other kinds of research. The kind of argument that is put forward may be considered in the form of an analogy. Suppose that we were studying the assets of a particular country; would we begin by making a painstaking and accurate inventory of every brick and stone present in that country? Or course not; a superficial survey of the buildings, highways, and so forth in the country would be all that is necessary, and we should devote our energy to other, more significant matters. However, the analogy is obviously a bad one, since the positions of individual bricks and stones are not crucial to how the country runs; we could remove and exchange stones and even buildings without any significant effect on the country's economy or politics. With the genome, on the other hand, it is an entirely different matter. A trifling change in a few base units can make all the difference between a person in good health and one afflicted with a crippling disease. To gain a full understanding in the long run, we do need the basic information that the Human Genome Project has revealed. But much more than that is necessary: we will need to know how the individual genes are related to the details of the genome, and the relationship between the genes and the human condition. On all of these matters, much has been discovered, but a great deal more research remains to be done.

One result that complicates the understanding of genetic problems is that the human genome, with its 3 billion units, does not correspond in any simple way to the picture that has emerged from studies of family relationships. Sorting out the relationship between the genes and the genome on the one hand, and the relationship between the genes and the human condition on the other, is obviously a matter of great complexity.

The manner in which characteristics are passed on from one generation to the next is only now beginning to be understood. It would simplify matters if the traits handed down from generation to generation were controlled by single genes, but this is rarely the case. It is true for some diseases, however, such as cystic fibrosis and muscular dystrophy, and in some cases the genes responsible have been identified. Attempts are therefore being made to develop techniques for repairing the damaged genes.

However, there is usually not a simple one-to-one relationship between a gene and a physical characteristic. Consider, for example, the color of eyes. It used to be thought that this would be controlled by a single gene, but now it is found that perhaps four to six genes are involved. The same is true of the color of skin. Most characteristics in humans are controlled by several genes, and the interactions of genes with one another are also involved in a complex way. Mendel's experiments with peas depended on a much simpler form of inheritance than is found with higher animals.

The control of genetic disorders is obviously much more difficult when many genes contribute to the condition. This is the case, for example, with heart disease, cancer, and diabetes. With cancer, over one hundred genes are already known to be involved, and it is suspected that there are many more. It is hoped that it will soon become possible to halt the onset of genetic disorders such as these before they start, but the task is one of great technical difficulty.

One particular matter on which the science of molecular genetics has already been able to shed some light is the question of race. For centuries the popular opinion has been that there exist races of people, usually defined by their skin color and other physical characteristics, who at some prehistoric time had quite different origins. The human race was supposed to have been divided into biologically pure lineages that originally were distinct from one another but that over the centuries have become somewhat blended by interbreeding. Religious respectability has sometimes been given to such ideas, for example by the suggestion that the "black," "white," and "yellow" races descended from Ham, Shem, and Japhet, the three sons of Noah. There has always been a tendency for people of one race to despise those of another race, and this has had a disastrous social impact. In many parts of the world, even in parts of the United States until a few decades ago, strictly enforced laws kept white and black "races" apart.

Racist attitudes are evident in the characterization of the hereditary condition known as Down syndrome. Because of a chromosomal error, the sufferers of Down syndrome are of diminished intelligence and have an unusual facial appearance. English physician J. H. Langdon Down (1828–96), who discovered the condition in 1866, called it Mongolism, believing that the individuals had somehow slipped down the evolutionary ladder so as to resemble the Mongols of central Asia, who were considered to be a lower form of life. It is amusing to note that apparently in Japan the condition was denoted by a word that means "Englishism." Modern English dictionaries say that the word Mongolism is offensive; I do not know if the Japanese dictionaries make a corresponding comment about Englishism.

Modern genetic research does not support the idea that there is a hierarchy of human beings, with white people of high pedigree of course at the top. It turns out that the visible variations between different groups of people, such as color of skin, are due to differences in a small group of genes, corresponding to only a small fraction of the total number. The genes determining color do not appear to be linked to a significant number of other genes, including those relating to intelligence. There is thus no reason at all for the misguided belief, for example, that black people differ in intelligence from white people. From many years of teaching large groups of students of many so-called races, I had myself, without making any scientific study of the question, come to the conclusion that there was no correlation between how good they were as students and the color of their skin. It is nice to learn that this anecdotal evidence seems to be supported by the latest genetic research.

If we take a group of white people from different countries and another similar group of black people we find that statistically the genetic differences between a randomly selected white person and a black person are only slightly greater than those between two white people or two black people. Put differently, the genetic differences between two white people from two European countries will be not much less—perhaps only 5 or 10 percent—than those between a white person and a black person. The differences between the majority of people in a country and the visible minorities thus lie largely in the visible characteristics and do not reflect anything more important. We are all cousins under our skin.

▲▲▲

Cells contain many different subcompartments called organelles. Significant information has been obtained about the *mitochondria* (one type of organelle). These are tiny, lozenge-shaped structures, present in the thousands in each of the cells of our bodies. Mitochondria are of enormous importance in our metabolism, since they bring about many chemical reactions essential to life. The interesting point about the mitochondria in humans is that they are transmitted by the mother only. A useful consequence of this is that it is now possible to make identifications of persons through the female line. Genetic tests of this kind established the identity of the remains of some members of the family of Nicholas II, the last emperor of Russia. Because mitochondrial genes evolve more rapidly than other genes, they are often used in the study of the evolutionary process.

Through similar studies scientists have concluded that the latest common ancestor of all humans *along the female line* is a woman, sometimes called Mitochondrial Eve, who lived a few hundred million years ago. On the basis of human remains found in Africa, it seems likely that Mitochondrial Eve lived on that continent.

▲▲▲

Work in molecular genetics has had an important practical consequence, the development of the field of genetic engineering. This is a vast field that continues to expand and has many applications. It is of particular significance to farmers: much disease on farms, for example, can now be controlled. Crops can be inoculated with artificially produced viral genes that limit the ability of invading viruses to replicate within plant cells. Fungal-resistant genes can be injected into clusters of corn cells, which are stimulated to replicate and grow into complete plants that grow much more satisfactorily than the original corn. There are many more examples of useful techniques of this kind. No doubt these new techniques will continue to play an important role in the difficult task of feeding the peoples of the world, so many of whom are now in great need.

Genetic engineering is also being used widely in medicine. It is employed, for example, in the production of many drugs that are difficult and expensive to extract from natural substances or to produce by ordinary chemical means. There are three important advantages to this method. One is that the procedures are usually simpler and cheaper to carry out. The second is that drugs produced in this way are usually purer than those produced chemically or those extracted from biological material. The third reason is that human genes can be used in their production, so the drugs are more compatible when used on humans. A good example is insulin, used to treat diabetes; animal insulin causes allergic reactions in some people, but insulin produced using human genes has no such disadvantage. Another example is provided by penicillin. The original strain of *Penicillium notatum* yielded only minute traces of penicillin from enormous vats of fungus. After years of painstaking work, biotechnicians were able to alter the genetic character of the bacterium, eventually producing much larger yields of the drug. The work involved procedures such as bringing about mutations by radiation. The results were highly successful, and if this genetic engineering had not been done few people could have been treated by penicillin, which would be extremely expensive; now the supply is adequate.

The story of the commercial production of the drug interferon is particularly interesting and significant. It was discovered in 1957 that interferon is produced by cells in the human body in response to the attack of viruses. The substance acts by producing a protein that stimulates the immune system, in this way inhibiting the spread of infection. Its importance was recognized at once, but the substance could not at first be marketed because of its scarcity and high cost. Many thousand human donors would be required to provide enough of the drug to be useful for a single patient, and a single dose would have cost perhaps fifty thousand dollars. The situation changed dramatically with the advent of genetic engineering in the 1970s: in 1980 a gene for human interferon was introduced into bacteria, the first time this had been done for a human gene. By cloning millions of bacterial cells from the original one, it was possible to produce a cheap supply of the drug, which is now widely used not only to combat viral infection, including the common cold, but also cancer.

Recently there has been much progress in the treatment of genetic diseases such as cystic fibrosis and muscular dystrophy by gene replacement. For these two diseases a single gene is involved and has been located. Various techniques are being used to insert a corrected version of a gene into a person's body. This is a field in which much research is being done and on which much progress is to be expected in the first decades of the third millennium. There has been criticism of the use of such techniques, which some think is unduly interfering with nature. Most would agree that creating "designer babies" by genetic techniques is undesirable, but there should be no moral objection to the treatment of genetic disorders. It would seem to me that, on the contrary, it would be immoral to fail to cure such diseases if it should become technically possible to do so.

▲▲▲

The theory of evolution is now accepted by all scientists, who regard its validity as being as sound as that of other great scientific theories such as the reality of atoms and molecules, the quantum theory, and the theory of relativity. It is not, however, so universally accepted by the public, because some people think that it conflicts with their religious beliefs. The evidence for evolution is so strong that the theory is as close to the truth as we can ever be in science. Let me summarize the main evidence for the theory.

Scientists do not claim to reach any absolute truth; they are merely concerned with what *works*. They say that a theory is true if it explains all the known facts obtained from observation and experiment, and if it can be used successfully to predict what will happen under another set of conditions. Darwin himself arrived at his theory of evolution because he knew a great many geological and biological facts, drawn during his five years of observation on HMS *Beagle*. During subsequent years a vast number of new relevant facts have been discovered. They have required a few modifications of Darwin's original theory, as almost always happens when a new theory is proposed, but everyone who is familiar with the subject is satisfied that the theory must be correct in its general outline.

An even more compelling reason for believing the truth of the theory is that the hard science on DNA, after its structure was determined by Crick and Watson, and on other genetic materials has all been completely consistent with evolution theory. We now know that the information that tells a cell how to function is coded in certain specific chemical units along a DNA molecule. The order in which these units occur has now been determined. The genetic code, by means of which the information in the genes is used in the synthesis of proteins, is now known precisely. What is remarkable is that the same genetic code applies to all bacteria, plants, and animals that have been studied. This points strongly to the conclusion that bacteria, plants, and animals have a common origin; we can calculate the probability that this identity could have occurred by chance, but it is exceedingly small. Another point is that long stretches of DNA have no function and are clearly the relics of the evolutionary process.

Another strong argument in favor of evolution is that there are close resemblances between the arrangement of the genes in species that are known from general biological observations to be closely related. Humans and chimpanzees, for example, are closely related, and their genetic makeup differs in only minor details; 98 percent of the genes of a chimpanzee are identical to those of a human. We have 46 chromosomes (in 23 pairs), whereas chimpanzees have 48, and the sequences of codons in the chromosomes are quite similar. There are greater differences for distantly related species, such as humans and mosquitoes (which have only 6 chromosomes) and where there are greater differences in the sequences of codons. All the genetic information of this kind that has been obtained is entirely consistent with Darwin's basic theory. Incidentally, we should not think that the number of chromosomes is any indication of status in the evolutionary scale. It is true that whereas we have 23 pairs of chromosomes, house mice have only 20. Chimpanzees, however, have 24 pairs and are not generally regarded as quite as intelligent as we are. Potatoes, which are surely not as intelligent as humans, also have 24.

Finally, biologists who work with bacteria and viruses are finding that they are evolving all the time. The influenza virus, for example, undergoes mutations in order to evolve into forms that are more resistant to antibiotics. Here is evolution occurring in front of our noses.

These details should make it clear that the theory of evolution now has as firm a basis as any other scientific theory, such as the atomic theory and the quantum theory. It is sometimes argued that the theory has no more validity than the ideas expressed in religious writings that are collectively referred to as "creationism," but such an assertion is entirely without foundation. The theory of evolution has many practical consequences, whereas creationism is completely sterile, leading to no further insight into nature.

▲▲▲

It seems that life appeared at least 3.5 billion years ago, about a billion years after Earth itself was formed. The evidence for this statement is that ancient remains of living organisms have been discovered, the age of which can be obtained by the important technique of *radioactive dating*. Various methods are used, all leading to the same answers.

One method is to use potassium-40, which decays into argon and has a half-life of 1.3 billion years. The potassium-argon ratio method is often used to date early fossil remains. Another method, using radioactive carbon, is particularly useful for estimating the age of more recent material, such as wood and bones; this technique is known as *carbon dating*, as noted earlier. Carbon-14 has a half-life of 5,730 years and decays to nitrogen-14 and a negatively charged beta particle:

$$^{14}_{6}C \rightarrow {}^{14}_{7}N + {}^{0}_{-1}\beta.$$

This carbon-14 isotope hardly exists at all in the planet's crust, but it is constantly produced in tiny amounts in the atmosphere by the action of cosmic radiation on the nitrogen molecules. It is the neutrons in the cosmic rays that convert nitrogen into carbon and hydrogen, as follows:

$$^{1}_{0}n + {}^{14}_{7}N \rightarrow {}^{14}_{6}C + {}^{1}_{1}H.$$

The action of sunlight is to cause some carbon-14 to combine with oxygen in the air to form carbon dioxide ($CO_2$), which enters into all living organisms. As long as the organism is alive, the amount of carbon-14 in it remains the same, being kept in constant supply by the fact that atmospheric carbon-14 can enter the system freely. When the organism dies, carbon-14 ceases to enter into the organism, and the carbon-14 already present begins to decay. The amount of carbon-14 that remains, for example in a fossil, thus enables scientists to determine how long ago the organism died.

The oldest fossilized organisms discovered so far were in rock formations in Western Australia. In the Pilbara area, about 40 km (25 mi) west of the small town of Marble Bar, John Dunlop, a geology student, discovered some stromatolites in 1980. These are structures that are deposited layer by layer by cyanobacteria, a form of bacteria that, like green plants, bring about photosynthesis, the process in which water and carbon dioxide are converted into organic compounds. Radioactive dating gave the age of these bacterial remains to be 3.5 billion years. Life had thus appeared at that early time, and photosynthesis, a somewhat complex process, had already evolved. Living systems must therefore have originated earlier than 3.5 billion years ago.

Subsequent to Dunlop's original discovery in 1980, many other signs of very early life have been detected. It is of interest that some of these were found near Darwin, in Western Australia, a town named after Charles Darwin, who carried out work in that region during his trip on the *Beagle*. He had been surprised to find no fossils. It now turns out that they are there but are too small to have been detected by earlier observers like Darwin, who lacked the advantage of techniques available to biologists today. In 1993, for example, J. William Schopf discovered in Western Australia some fossils of single-celled organisms that lived on Earth 3.5 billion years ago.

In the 1990s evidence for early life was also found in a remote mountainous region at the edge of the massive Greenland ice sheet. An expedition led by Gustav Arrhenius of the Scripps Institute of Oceanography in California discovered traces of carbon-containing deposits in the form of tiny grains laid down by primitive living organisms. The evidence that they were living comes from a particularly sensitive radioactivity technique, developed more recently. We have seen that carbon-12 is the most common isotope of carbon, the nucleus of its atom containing six protons and six neutrons. A small proportion (about 1.1 percent) of the carbon in nature occurs as carbon-13, in which there is an extra neutron in the nucleus. Because of its smaller mass, carbon-12 reacts chemically slightly faster than carbon-13, but otherwise the chemical properties of the two forms are almost identical. As a result of this slightly greater rate of reaction, carbon-12 accumulates rather more in living organisms than carbon-13 does, and the excess gives an indication of the age. In the Greenland rocks, the carbon-12 was about 1 percent higher than normal, which is confirmation that the material was really of biological origin. The age was found to be about 3.85 billion years.

It is impossible to be certain how the earliest living organisms came into being. Various suggestions have been put forward. At one time it was thought that protein molecules play the essential role in living systems, and various attempts were made to see whether, in the kind of atmosphere believed to exist on Earth in those early times, they might have been formed naturally from available substances. In 1953 experiments were carried out by Stanley Miller (b. 1930), a graduate student in the laboratories of Harold Clayton Urey (1893–1981), who had won the 1934 Nobel Prize for chemistry and was distinguished for his work on isotopes and the distribution of chemical elements. Miller, with Urey's rather grudging approval at first, made mixtures of simple chemicals thought to be present on Earth 3.5 billion years ago and subjected them to electric discharges, which simulated lightning. He found, for example, that if water vapor, hydrogen, methane, and ammonia were held at 100 degrees Celsius (212 degrees Fahrenheit), and subjected to electric discharges for a week, some amino acids and other, more complex substances were formed. Since protein molecules are produced by the combination of amino acids, these experiments seemed to make plausible the possibility that the first life forms had arisen in this way.

We now know, however, as a result of the determination of the structure of DNA by Crick and Watson and by later studies, that the key substances in living systems are not proteins but are DNA molecules, which are directly concerned in replication. The formation of DNA is much less likely to occur by chance than the formation of protein, since DNA is a much more complicated molecule than a protein. Of course, the production of a molecule of DNA, which must have happened at some stage, may not have been the first self-replicating substance to be formed. Graham Cairns-Smith of the University of Glasgow has suggested that inorganic clay crystals may have been the first replicators, DNA coming at a later stage of evolution.

It seems impossible for us to be certain about how life originated on Earth, but it is interesting to consider whether there is a reasonable probability of its happening by chance. Several facts have to be taken into account. One is that life apparently began at least 3.5 billion years ago, which is about a billion years after our Earth had cooled enough to make life at all possible. Another is that astronomers have estimated that there are about 100 billion billion ($10^{20}$) planets, some of which could sustain life. If life had not begun on this planet it probably would have begun on one of the many others, so we would then all be occupying another planet. Even taking this into account, however, most scientists consider it highly improbable that molecules as complex as DNA could have been formed by accident. Perhaps simpler molecules, such as those considered by Cairns-Smith, were formed in this way, the DNA molecules emerging at a later stage of evolution. It is difficult to quarrel with the view expressed many years ago by eminent astronomer Sir James Jeans (1877–1946) that the universe does not seem to have been created with the purpose of sustaining life, most of it being highly inhospitable to life.[4] A tiny minority of scientists do, on the other hand, sometimes for religious reasons, subscribe to the *anthropic principle,* according to which the universe has been deliberately constructed in such a way as to lead to the evolution of human life. I personally see no evidence for such a point of view.

It is usually assumed that our ultimate ancestor originated on Earth, but we should consider whether life might have begun elsewhere and was brought to Earth, perhaps by a meteorite. The possibility that it originated on Mars is explored in a most interesting way by Paul Davies in his book *The Fifth Miracle: The Search for the Origin and Meaning of Life.*[5] Whereas Mars is now inhospitable to any form of life, there is evidence that at one time it had lakes and rivers as well as an atmosphere more like our own. It is therefore possible that very primitive life forms began on Mars and were conveyed to Earth by a meteorite. Small samples brought back from Mars to Earth in 1997 were said by some to indicate the possibility of life but, as Davies makes clear, the evidence is far from conclusive.

In whatever way the initial replicating molecule was formed, we can hardly doubt, from the evidence, that all living things now on Earth are descendants of a microorganism formed about 4 billion years ago. One reason for believing this relates to the genetic code and the laws of probability. We have seen that we can think of the genetic code as a tiny dictionary in which there are 64 codons (the number of possible ways in which the four bases T, C, A, and G can be arranged in triplets). Each gene sequence ends with a "stop codon," which indicates the end of the gene (sometimes referred to as a punctuation mark). These 64 codons are the codes for the 20 amino acids and the punctuation mark; it is as if the 64 words of one very primitive language were translated into an even more primitive one in which there are only 21 words. The odds of getting the same 64:21 coding twice by chance can be calculated as less than 1 in $10^{30}$ (a million million million million million). It is thus highly unlikely that the different species of bacteria, plants, and animals were created separately.

Another good reason for thinking that all living things have a common ancestor is that all of the molecules that play essential roles in living systems have the same *chirality*, or "handedness." We can easily understand this by looking at our two hands. They are different from each other but are similar in the particular sense that one hand is more or less the same as the mirror image of the other. The same is true of the amino acids. A given amino acid can exist in either of two forms, the D form or the L form; each is the same as the mirror image of the other, but because of a particular feature in their structures they are not identical, in that one cannot be superimposed on the other. It turns out that in every living system that has ever been studied, all of the amino acids are in the L form. Moreover, the D forms cannot be accommodated into living systems. If, for example, someone gave us a piece of meat that had been synthesized with every amino acid in its protein molecules in their D forms, it might be made to look exactly the same as normal meat. If we ate it, however, the enzymes in our stomach and intestines would be quite unable to digest it.

The fact that all organic molecules in living systems, including the proteins, the DNA, and the RNA, contain only L molecules, is another strong piece of evidence for a common origin. A form of life using all D molecules, or containing various mixtures of L and D forms, would have been just as possible, but has never been observed. Perhaps life did originate several times and in different ways, but if so only one arrangement, all L, has survived to the present time. If living things today came from several origins, surely some of them would contain D rather than L molecules.

Over the ages, living organisms have adapted to their environment to increase their potential for survival and reproduction. As certain characteristics are transmitted from parent to offspring, and as mutations occur from time to time, some organisms prove to be poorly adapted to their environment and die out, while others increase in number. In accordance with Darwin's theory, this is survival of the fittest: the weakest organisms (from the point of view of their ability to live long enough to reproduce) fail to survive, while the stronger ones continue the line.

Different species can be formed when mutations occur. They can also be formed when organisms of the same species become separated geographically. The separated species then develop in different ways, becoming distinct from each other after many generations as a result of natural selection until finally they are so different as to be unable to interbreed. By definition they are then separate species. For example, why did red squirrels and gray squirrels, which resemble one another in so many ways, become different species, in the sense that they cannot interbreed? Red squirrels are indigenous to Europe; the few that are seen in North America have been transported there by humans. Gray squirrels, sometimes called Canadian squirrels, are indigenous to North America, although some were introduced by humans into Europe, where they propagated rapidly and are now numerous. From their great similarity, we can conclude that long ago the ancestors of the red and gray squirrels belonged to one and the same species. It seems likely that many generations ago a group of squirrels became separated into two populations, perhaps by a mountain range or perhaps by water. Later, as Earth's configuration changed, they were separated by the Atlantic Ocean. They continued to breed, and over the generations, as a result of mutations and natural selection, the members of the two populations became sufficiently different from one another that they could no longer interbreed.

There are three main classes of living beings: bacteria, plants, and animals. In the early days of the theory of evolution, before much had been learned about genetic mechanisms, it was thought that evolution occurred by simple pathways, something like the following:

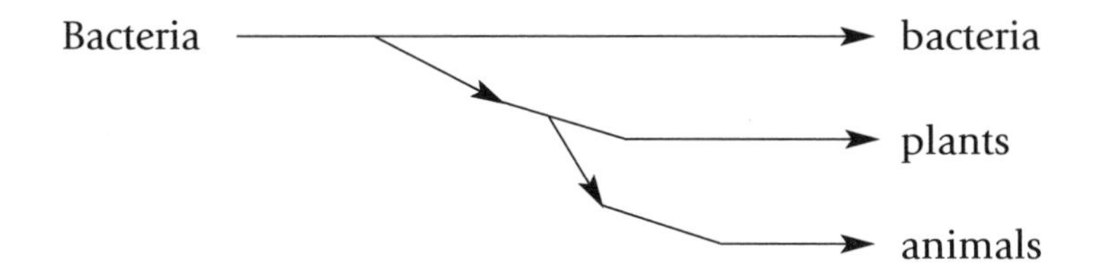


Once the first animals were formed, it was thought that there was a progression through monkeys, chimpanzees, Neanderthal man, and finally our species, *Homo sapiens*. We now know that this is wrong. We did not descend from monkeys or chimpanzees, nor even from Neanderthal man. They belong to different branches of what is often referred to as "the tree of life" (see fig. 57).

The details of this tree are still being studied and will continue to be studied for a long time, since the matter is very complex; after all, there are 30 million species to deal with. Modern molecular genetics has nonetheless contributed greatly, as shown by the following simple examples. There is much evidence from conventional biology that apes, gorillas, and chimpanzees are closely related to one another and to humans. Examination of their genetic material shows that the DNA differences among them amount to only 1 or 2 percent. It is inferred from this and other evidence that the common ancestor that we share with apes, gorillas, and chimpanzees existed 4 or 5 million years ago.

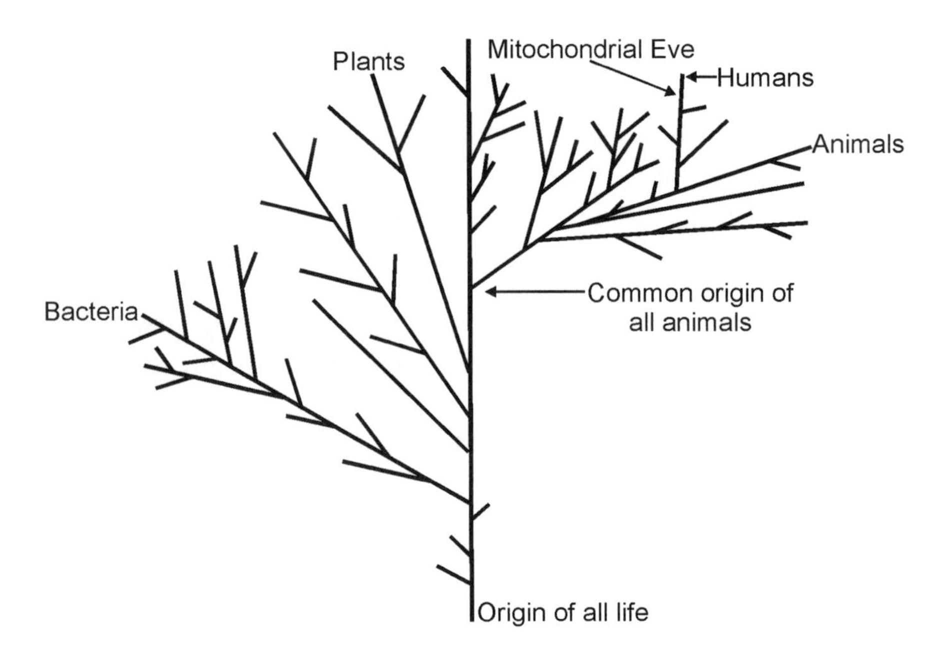


**Figure 57.** A very simplified tree of life. The present day is at the top. Some of the "branches" never reach the top and represent extinct species, such as dinosaurs and Neanderthal man. A more detailed (and enormously large) diagram would show 30 million "limbs." It would also show that for every limb that reaches the top, there are about ten that do not.

Since the rate of mutation of a gene appears to be approximately constant, we can estimate the time at which different species diverged from a common ancestor from the differences between the codons in the species. In all cases such conclusions have been consistent with other information leading to estimates of when new species emerged. For example, the essential protein called cytochrome c is made up of about one hundred amino acids, its formation in humans being determined by a string of 339 codons. A monkey, a close relative of ours, also has a cytochrome c that differs from ours by only one amino acid out of the hundred or so. In the monkey it has almost exactly the same properties and performs the same functions; the section in the monkey's genes that produces it differs from ours in only one codon out of the 339 codons. This gives overwhelming support to the idea that monkeys and humans came from a common ancestor, and it is estimated that they diverged into separate species about 4 or 5 million years ago.

On the other hand, when we compare the cytochrome c from yeast with that from humans, 45 out of the 339 codons are different. Again this makes sense, since humans are very different from yeast, and the branch leading to humans and yeast obviously split in a much more distant past—much more than 4 or 5 million years ago. The difference between the numbers for pigs and yeast is also 45, and this can be understood, since the branch leading to pigs split from the one leading to us much more recently than their common branch separated from the one leading to yeast. There are many relationships of this kind, all giving strong support to the theory of evolution. There has been some disagreement between experts about some of the fine details of evolution, but in view of the complexity of the situation and the vast number of genes involved, this is inevitable.

The general direction of evolution has been to develop, from a single species ("Origin of life" in fig. 57), more and more species as time has passed. This increase in diversity of species is analogous to increased disorder or, in the extreme, to chaos. This theme will be explored further in the next chapter.

10

# CHAOS THEORY

*Chaos* traditionally means "utter confusion," and the word is used in this sense in connection with the second law of thermodynamics, discussed earlier. We saw that processes occur in the direction of increasing entropy, which means increasing disorder, and we referred to an "arrow of time" corresponding to increasing disorder. And in the last chapter, we saw the analogy between evolution and disorder. This means that the universe is becoming more and more disordered, such that eventually complete chaos will prevail. However, our Sun may survive in its present form for another several billion years, and long after that, much will still be going on in the universe.

In the modern theory of chaos, the word *chaos* is used in a much more limited sense. It relates to certain occurrences for which the final outcome is impossible to predict. This kind of chaos is obviously very different from the complete and utter chaos that will exist at the end of time, and it is unfortunate that the same word has come to be used for it. Chaos in this much more limited sense is sometimes referred to as *noncatastrophic chaos*. It is also known as *deterministic chaos*.

Chaos theory is the science of the unexpected—in two senses: it deals with unexpected things, and its name is appropriately unexpected! One reason that this special use of the word *chaos* is regrettable is that it forces people to use expressions that, at least to me, seem rather comical. It is sometimes stated, for example, that our solar system is "slightly chaotic," by which is meant that the chaotic behavior is hardly detectable at the present time but might become evident in a million years or so. Saying "slightly chaotic" sounds to me as ludicrous as saying that something is slightly unique, or that someone is slightly insane, slightly destitute—or slightly dead.

Modern chaos theory deals with many of the familiar occurrences of everyday life. It is involved whenever it rains, whenever we turn on a tap, and whenever an epidemic strikes. A familiar example may help us to understand what chaos is in this sense. Imagine a lawn that becomes infested with beetles, and suppose that we do nothing about it. The lawn deteriorates, and the beetle colony grows rapidly. The lawn becomes so sparse that beetles start to die of starvation, and the lawn then starts to revive and looks more respectable; the few beetles still alive replicate, and the cycle starts over again. It is easy to see that the quality of the lawn may go through stages when it looks good and when it looks poor; the population of beetles also varies, from small (when the lawn looks good) to large (when the lawn looks poor).

The final outcome may be that the grass and the beetles all die. Another possibility is that the lawn might settle down into an equilibrium state, in which the beetle population remains fairly low and the lawn is not too bad but not at its best. Which would happen would be impossible to predict, even if there were no outside influences. This is true even if we ignore environmental factors such as the weather or the addition of chemicals, which would further influence the outcome. This is an example of what is called *predator-prey chaos*. It is also found, for example, with a population of cheetahs and antelopes, the cheetahs relying for their survival on a supply of antelopes as food, and the antelopes running away from them as fast as possible. The population of antelopes is then apt to fluctuate in a chaotic manner, and the population of the cheetahs will also fluctuate unpredictably.

Another example is a lake, initially completely free from external pollution, into which a small number of fish are introduced. At first they will reproduce rapidly but soon will have to compete with one another for food and will also suffer from the pollution that they themselves create. Their population may then decline and later recover. Under some circumstances the result may be a population that rises and falls in a completely unpredictable way.

Yet another familiar example relates to three well-known games —croquet, billiards, and curling. In each case one aspect of the game involves propelling a ball (or stone, in the case of curling) in such a way that it first hits one ball, bounces off it, and then hits another. In billiards this is called making a carom. It is relatively easy to hit the first ball, but it takes more skill to hit the second one. If hypothetically you were called on to hit successively a whole series of balls it would become increasingly difficult. Only the most skillful of players would have much chance of hitting a third ball, and hitting the fourth, and hitting subsequent balls would probably be a matter of luck. It is easy to see from this example what the essential feature of chaos is. We are dealing with a series of events, each one of which depends much more critically than the previous one on the precise initial conditions.

A pinball machine is helpful in showing us how probability applies to chaotic motion (see fig. 58). If the ball is released just above the top pin so that it hits the pin, it next hits either of the pins on the second row, and to simplify things we will assume that it hits either of them with equal probability. If it hits the left pin of row 2, for example, it has an equal chance of hitting pins 1 and 2 of row 3, and so on. There are ten rows in the machine shown in figure 58, and it is easy to calculate that there are exactly 512 (= $2^9$) trajectories from the top to the bottom row. Only one of these leads to the left-hand pin in the bottom row, so we know that the probability of hitting that pin is 1 in 512. The probability of hitting the right-hand pin in that row is also 1 in 512. There are higher probabilities of hitting the inner pins, which can be calculated. Consider just the fourth row, which contains four pins. There is just one chance of hitting the left-hand and right-hand pins in this row. As to the other two (the second and third), it is easy to see that there are three trajectories leading to them. There are eight ($2^3$) trajectories in all, so the chance

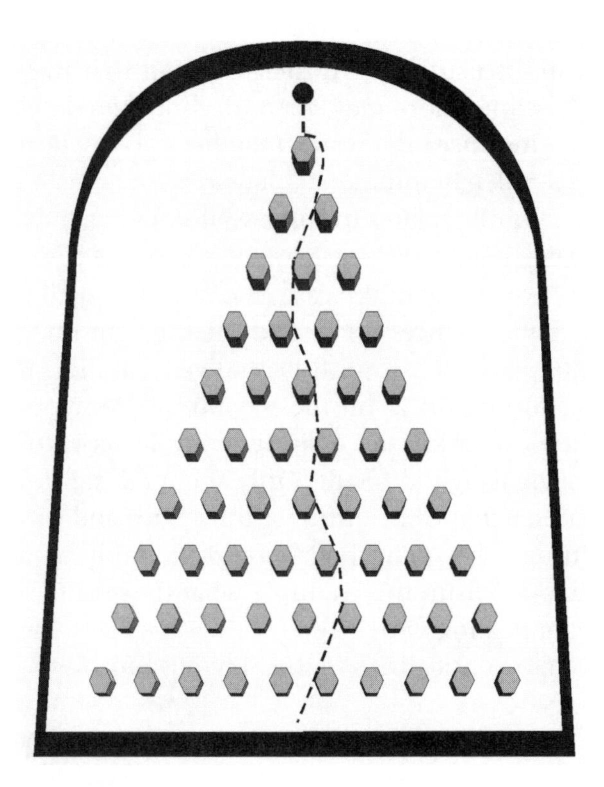

**Figure 58.** A simple pinball machine. The path of the ball is very sensitive to its position; it is therefore impossible to predict the outcome from the initial conditions. In other words, the behavior is chaotic. Where the ball finally lands, however, is not random since its probability can be calculated.

of hitting pins 1 or 4 is 1 in 8, whereas the chance of hitting 2 or 3 is 3 in 8.

The chance of hitting any one of the pins is determined by simple probability principles. This emphasizes the important distinction between "chaos" and "randomness." If the pins in the bottom row were hit completely randomly, there would be an equal probability of hitting each one of them, instead of a much lower probability of hitting the outer ones as compared with the inner ones. Thus, although the motion of the ball is chaotic as it rolls down the machine, the probability that it hits a certain pin is determined by the construction of the pinball machine. We can express this by saying that we have *deterministic chaos*. This is the basis of the methods used to determine whether behavior is random or chaotic; we apply probability tests to the data, and if probability relationships apply we know that we have chaos; otherwise we have randomness.

Noncatastrophic chaos is sometimes the outcome when there is oscillatory behavior. An example is provided by a heartbeat, which is controlled by natural pacemakers. Sometimes these do not work together properly, such that there are alternate long and short gaps between the beats. Under more extreme conditions the beating becomes highly irregular. In one serious condition, called *ventricular fibrillation,* the heart flutters erratically instead of expanding and contracting rhythmically when it pumps blood. A small change in the timing of one beat makes a bigger change in the timing of the next; the beating becomes chaotic.

There are many familiar processes that do not give rise to deterministic chaos. Examples are the swinging of a simple pendulum, ordinary engines such as those in automobiles, and most chemical reactions including those that occur in the body. For these we can set up and solve the appropriate mathematical equations; then, knowing the initial conditions, we can predict exactly and without any ambiguity just how the process will occur over a period of time. For such systems the behavior may be complex and difficult to work out, and we may have to use numerical methods and even a computer. Often the results present no surprises, in that what occurs is just what we would expect. This is fortunate in many practical situations; it would be disastrous if when we applied the brake to a car it would sometimes accelerate for no ascertainable reason.

Although chaos theory has developed recently, it deals with many things with which we are all familiar: a flag flapping in the wind, traffic jams on the highway, the progress of a war, and the rise and fall of the stock market or of an epidemic. What commonly happens when an epidemic, such as AIDS or SARS, develops is that after the first case is detected the number of cases increases rapidly. If appropriate preventive action is taken the number of active cases might later fall, but that fall may be followed by a rise. The prevalence (i.e., the number of cases per one hundred thousand people) of the disease in the population may in fact pass through several maxima (highs) and minima (lows), and the final outcome would be uncertain. If the preventive action taken is adequate, the oscillations may be followed by a decrease in the number of cases, and the disease may finally be eradicated, as in the case of SARS. Otherwise, as in the case of AIDS, the disease may get out of control. It is important for authorities to realize the possibility of oscillations in the intensity of an epidemic and not to relax efforts after the first encouraging sign that it might have been brought under control.

Chaos theory turns everyday happenings, previously avoided and sometimes scorned by scientists, into legitimate subjects of scientific study. The theory unites researchers in different fields and helps reduce the excessive specialization that is so prevalent today.

▲▲▲

Until the nineteenth century basic scientific theories were *deterministic,* meaning that the future was believed to be entirely determined by the past. This had been assumed to be true by early scientists such as Isaac Newton. A century after he did his work, French scientist Pierre Simon, Marquis de Laplace (1749–1827; see fig. 59) argued that if we could specify the exact state of the universe at a given time, we could in principle deduce its state at any future time. Laplace, who was born

**Figure 59.** Pierre Simon, Marquis de Laplace (1749–1827), French mathematician and astronomer, suggested a theory of the origin of the solar system. He was one of the first to put probability theory on a firm foundation. (Edgar Fahs Smith Collection, University of Pennsylvania Library)

in Normandy, became professor of mathematics at the École Militaire and was elected as a member of the Académie des Sciences at the early age of twenty-four. He was one of the most versatile and influential scientists of all time: he made valuable contributions to a wide variety of problems in mathematics and physics; he developed a highly productive philosophy of science; and he played an important part in establishing the modern scientific disciplines. He was interested in experimental science as well as in mathematics and entered into a collaboration with eminent French chemist Antoine Lavoisier (1743–94) on theoretical and experimental investigations of heat.

From 1768 onward Laplace worked on integral calculus, astronomy, cosmology, the theory of games, and the theory of probability. During the revolutionary period in France, from 1789 to 1805, Laplace's reputation was at its height. He was an active member of the commission that introduced the metric system in 1799, and he played a dominant role in the creation of the Institut de France and the École Polytechnique. His comprehensive *Traité de méchanique céleste* (Treatise on Celestial Mechanics) appeared from 1799 to 1825; the later volumes contained much of Laplace's physics with its "astronomical" theory of matter. Laplace presented a copy of the book to the emperor Napoleon, who commented that, although dealing with the universe, the book made no mention of its Creator. To this Laplace stiffly and bluntly replied: "Je n'avais pas besoin de cette hypothèse-là" (I had no need for that hypothesis). Napoleon, much amused, told the story to many people, including distinguished mathematician Joseph-Louis Lagrange (1736–1813), whose comment was, "Ah! c'est une belle hypothèse; ça explique beaucoup de choses!" (Ah! that is a nice hypothesis; it explains many things).

For a short time Laplace served as minister of the interior under Napoleon, who found him ineffective as an administrator; Napoleon said later that Laplace "saw no question from its true point of view; he saw subtleties everywhere, had only doubtful ideas, and finally carried the spirit of the infinitely small into the management of ideas." After six weeks as minister, Laplace was dismissed by Napoleon, who however continued to hold him in high esteem for his scientific achievements and later appointed him to the senate. When in 1814 the monarchy was restored in France, Laplace hastened to offer his services to the Bourbons and was rewarded with the title of marquis. Though an excellent scientist, he was politically a vain, self-seeking, calculating opportunist who shifted his political opinions as occasion demanded, always managing to retain a position of influence.

The first important challenge to determinism came from Maxwell's interpretation of the second law of thermodynamics in terms of probability. This interpretation showed that events do not always follow from one another by strictly deterministic laws, but sometimes as a matter of pure chance. Another kind of challenge came at about the same time from French mathematician, physicist, and philosopher Jules-Henri Poincaré (1854–1912; see fig. 60). He was born in Nancy, France, the son of a physician, and at the early age of twenty-seven was appointed professor of mathematics at the University of Paris, where he remained for the rest of his life. During his career he contributed to a vast range of topics in mathematics and applied science. In spite of this he was singularly absent-minded and clumsy, being quite inept with the simplest of mathematics, including arithmetic.

**Figure 60.** Jules-Henri Poincaré (1854–1912), French mathematician and philosopher, made important contributions in many areas of pure and applied mathematics. (Edgar Fahs Smith Collection, University of Pennsylvania Library)

Poincaré's challenge to determinism came as a result of his mathematical attack on what is called the three-body problem. An example of this is the Sun-Earth-Moon system, the mathematics of which Newton had attacked two centuries earlier. Newton successfully solved the dynamics of a planet such as Earth moving in its orbit around the Sun. He then attacked the problem of Earth moving around the Sun considering also the Moon moving in its orbit around Earth. At first sight one might think that that such a three-body system would be solvable by use of the same procedures, but this is by no means the case. Newton could not find an exact solution to the dynamical equations, in the form of equations that described the motions of the three bodies. It is now known that there is no such solution.

Newton tried to find an approximate solution to the problem by means of a technique that mathematicians refer to as the "perturbation method." The principle involved is to start with the solution for the Sun-Earth problem, which is the dominant problem. Newton then added on the influence of the Moon as a perturbation to this solution. This proved exceedingly difficult, and he worked on it persistently for a full year. He finally obtained a solution but was disappointed to find that it did not give the orbit of the Moon as accurately as he had wished. Newton always considered his work on the Moon to be his great failure, later lamenting, "Never did I get bigger headaches than when I was working on the problem of the Moon."

Subsequent to Newton's work on the three-body problem, several other eminent mathematicians also tackled it, but with little success. They could not overcome the fundamental difficulty that the perturbation introduced by the Moon was just too big to lead to satisfactory results. Poincaré's success came as a result of introducing a completely new method, one that is used to this day. Instead of using the ordinary three-dimensional space with which we are familiar, he used *phase space*. The conventional three dimensions are dimensions of position only. We can, however, extend this to three additional dimensions relating to velocity. The motion of a ball that has been thrown into the air can, for example, be described by three position dimensions and three dimensions of velocity, by which we mean the components of the velocity along three axes.

Poincaré began his phase-space attack on the problem as a result of a mathematics contest organized in 1889 by the University of Stockholm to mark the sixtieth birthday of Oscar II, king of Sweden and Norway (which were then one country). The problem posed in the contest was to determine the stability of the solar system. One possibility was that the solar system was stable, in the sense that the planets would remain in their precise orbits forever, always retracing the same paths. Alternatively, the orbits of the planets, as a result of the gravitational attractions of other planets, might change radically in the distant future. Previous mathematical treatments had led to conflicting conclusions, and the purpose of the contest was to come to a satisfying decision.

Having introduced the phase-space technique, Poincaré, aged thirty-five at the time, felt in a good position to enter the Stockholm contest. In doing so he was led to the conclusion that in certain complex dynamical situations such as the three-body problem, a tiny change in the initial position or velocity of one of the three bodies could lead to a complete change in its orbit. A tiny change, in other words, could lead from order to chaos. Regularity and chaos, he found, are not completely separate situations but are closely interrelated, the unpredictable being never far away from the predictable. Thus, what seemed a fairly simple system such as the Sun-Earth-Moon system, although precisely governed by Newton's gravitational laws, could give rise to unpredictable behavior. The deterministic universe of Newton and Laplace was therefore not a completely correct one; unexpected behavior may occur.

Poincaré won the contest of the University of Sweden, and his published paper constitutes an important explanation of the basic ideas of chaos theory.[1] In his book *Science et méthode,* published in 1909, he summarized his main conclusion: "There are situations where small differences in the initial conditions can produce very large ones in the final result: a small error in the former can lead to a huge error in the latter. In those cases, predictions become impossible." In spite of this very clear statement of the theory, few people at the time paid any attention to it, and the idea was not revived until the latter half of the twentieth century.

Earlier we saw that another kind of challenge to determinism came in the 1920s with the advent of quantum mechanics, and, in particular, with Heisenberg's formulation of the uncertainty principle. According to this principle it is impossible to determine the exact position and momentum of a particle at the same time. As a result, we cannot predict the exact course of events but can sometimes make estimates of probabilities.

Noncatastrophic chaos theory, however, relates to a *deterministic* type of uncertainty, which would occur even if there were no restriction imposed by the Heisenberg principle. Even if the uncertainty principle did not affect the situation, we still could not always predict what is going to occur. The reason appears when one analyzes the situation mathematically. With certain kinds of dynamical equations there is a basic instability in the behavior. Computer calculations carried out with almost exactly the same starting conditions lead to different outcomes if they are carried out several times.

▲▲▲

Although Poincaré had deduced the essential aspect of chaos theory before the end of the nineteenth century, almost no attention was paid to it until the middle of the twentieth century. Revival of interest in the theory was brought about to a considerable extent by the development of computers, which enabled numerical calculations to be made much more rapidly than before. The laborious calculations on the Sun-Earth-Moon system that Newton carried out over a period of about a year can be made today in a few seconds, allowing for much more extensive calculations to be made.

An example of this with which we are all familiar relates to weather forecasting. We know from practical experience that we can usually rely on weather forecasts for the next three or four days but that those made for a week or so ahead can be completely wrong. There was a striking example of this in the United Kingdom in 1987. On October 15 of that year there was a serious storm that did considerable damage; in Kew Gardens in London, for example, many trees were destroyed. Storms of hurricane dimensions, of which this was one, are rare in the United Kingdom, where it is commonly said that hurricanes never occur. What created concern at the time was that the weather forecasts had given absolutely no warning of the storm, and the public demanded an explanation; newspaper headlines screamed, "WHY WEREN'T WE WARNED?"

The meteorological office that had failed to give the warning used a computer that operated at a speed of 400 megaflop, which means that it performed 400 million calculations a second. The word *flop* is an acronym for *fl*oating point *op*erations per second, and since the word *flop* is used informally for something that fails, this computer usage invites sarcastic humor; megaflop = big failure. A few days before October 15, 1987, the weather office had correctly predicted a bad storm but one that was heading not for England but straight for the middle of the North Sea. To everyone's surprise, it suddenly changed direction. This is essentially the problem with weather forecasting: if all goes well the forecasts can be excellent, but unexpected things can happen. As a weather expert is reported to have said, "Today we can predict the weather accurately—provided that nothing unexpected happens." I remember reading a newspaper story years ago about a large group of distinguished meteorological experts who were holding an international conference in Oxford; when they left the building for lunch it was pouring rain, but no one had a raincoat or umbrella.

The fault, of course, does not lie with the meteorologists or their computers. No doubt they do their work admirably for the most part, but they are sometimes inevitably defeated by the fact that weather is a potentially chaotic phenomenon. This was clearly demonstrated by some computer investigations carried out in the 1960s by American meteorologist Edward Lorenz (b. 1917) of the Massachusetts Institute of Technology. In order to study weather patterns he devised highly simplified differential equations to represent meteorological processes such as the movement of air and the evaporation of water. His aim was to make predictions of such phenomena as the temperature, the direction of the wind, and the onset of rain or snow at later times. He introduced into his computer programs certain initial conditions and let the computer run in order to predict the values of the various meteorological parameters after various periods of time. In one case he repeated some calculations from a time at which the computer had given the number 0.145237 for a particular quantity and in doing so decided to round off the value to 0.145, thinking that such a small difference would be of no consequence. To his great surprise he found, even with his highly simplified model, that the second calculations led to predictions that were much the same for the first few days, but after that there were enormous divergences, no similarity between the predictions remaining after more than a few days (see fig. 61). At first he assumed that the computer was malfunctioning, but much further work showed the effect to be real.

This means, of course, that there are fundamental and unavoidable limits to the ability to predict the weather. There are difficulties even when the mathematical equations have been greatly simplified, but the difficulties are much greater for the conditions actually existing. The computers can make reliable predictions for the next

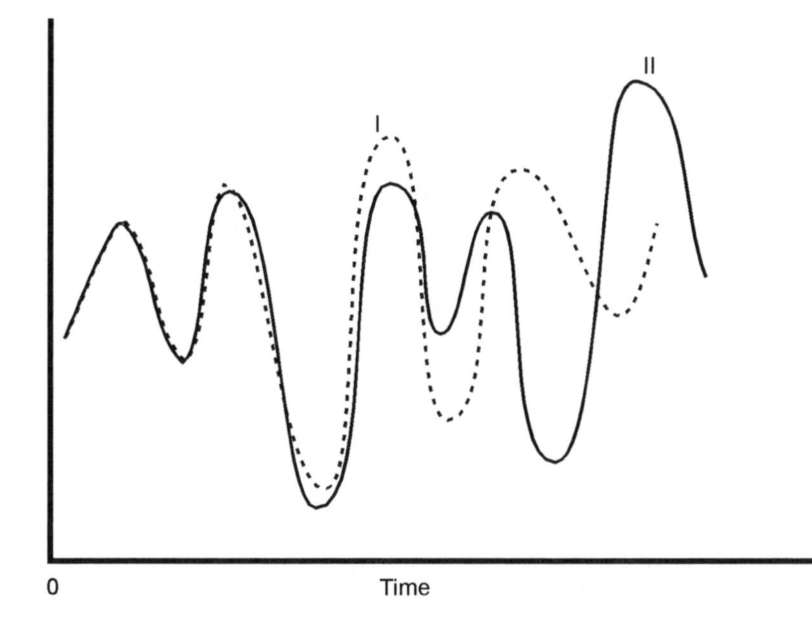


**Figure 61.** A weather parameter, such as temperature or wind velocity, plotted against time, as predicted by a computer for two almost identical initial conditions (curves I and II). Edward Lorenz found that the curves were more or less identical for the first two or three days but that they later diverged more and more, such that it was impossible to make reliable predictions for more than a few days.

few days, but reliable predictions are increasingly impossible for later times. In 1972 Lorenz suggested what is called the "butterfly effect," with his paper titled "Does the Flap of a Butterfly's Wings in Brazil Set Off a Tornado in Texas?"

▲▲▲

The conditions under which chaos can develop have been worked out mathematically. One is that at least three individual processes must be involved; this condition is usually satisfied. Another condition is that at least one of the processes should be such that the effect is not directly proportional to the cause but depends upon it more strongly; in mathematical language we say that the behavior is nonlinear. Another condition is that there must be what is called *positive feedback.*

We discussed feedback earlier, where we saw that negative feedback occurs with the thermostats that control the temperature in our homes; it also controls the temperatures of stars. Positive feedback is illustrated by a thermostat that has been wired incorrectly such that if the temperature is too low, the heating system turns off and if it is too high, it turns on. If, then, the temperature happens to be too high, the furnace makes it still higher, and the temperature eventually settles at some high value, corresponding to the maximum performance of the furnace. Positive feedback also happens in a chemical or nuclear explosion; a nuclear explosion; for example, occurs when uranium-235 reaches a critical mass.

A familiar example of positive feedback that many of us have experienced firsthand is at an event such as a party. What often happens is that the sound level soon becomes so high that in order to communicate we have to shout at the top of our voices. It would obviously make more sense if everyone agreed to speak in a normal voice, but that would only work if police were assigned to haul away offenders (negative feedback)—probably not the most popular way to throw a party. With a prior agreement but no police we can be sure that someone would soon speak louder, forcing others to do the same, and the sound would soon reach a high level.

In all the examples of chaos we have considered, it is easy to see that there is feedback. In the case of the beetles affecting the growth of grass, the number of beetles affects (adversely) the growth of the grass, whereas the amount of grass affects (favorably) the growth of the beetle colony; we have negative feedback in one direction and positive in the other, and it is the positive feedback that is essential for there to be chaos.

▲▲▲

Chaos theory helps us to understand the evolution of the universe and of life in it. The formation of the galaxies, with their complicated stellar and planetary systems, seems to imply the creation of order out of disorder, and we might think that it violates the second law of thermodynamics. The same difficulty seemed to arise for the formation of complex structures such as the eye, found in the higher life forms; how, it has been argued, can such complicated organization arise from much more disordered states? Snowflakes have beautifully complex structures; how can so many myriads of them descend upon us from highly disorganized storm clouds without there being a violation of the second law? Chaos theory provides the answer. The second law applies only to the final state of a process; intermediate states can show oscillations or complex types of behavior such as are sometimes found in experiments on complex chemical reactions.

Further understanding of this problem is provided by the work of mathematical physicist Benoit Mandelbrot (b. 1924) on what he called *fractals*. Of Polish-French origin, Mandelbrot was educated at the École Polytechnique in Paris and later visited the United States as an IBM Research Fellow, later working at Yale University. He worked on a wide variety of problems in applied mathematics, including mathematical linguistics, game theory, and economics. The contribution for which he is best known resulted from his work on the problem of noise on telephone wires used for computer communications. It was assumed that this noise would occur randomly, but when he studied it on shorter and shorter time scales he found that the distribution of the noise spikes continued to have the same characteristics as the original.

This led him to the realization that certain geometrical patterns also have this same characteristic. A snowflake and a cauliflower are familiar examples, in that when a tiny region of them is greatly magnified, the result looks much the same as the original. A coastline is another example of a naturally occurring fractal. A map of a coastline shows a certain distribution of bays and promontories. If, for example, we consider one particular bay and magnify it greatly, the result will show a similar distribution of bays and promontories; each one of these on magnification reveals a similar pattern, and so on *ad infinitum*. This work led Mandelbrot to his famous paper "How Long Is the Coast of Britain?" in which he showed that the answer depends on the scale with which it is measured: the shorter the measuring rod, the longer the measured coastline. His important conclusion was that as we reduce the length of the measuring rod the measured length of the coastline does not approach a finite value, as generally expected; instead, it approaches infinity.

When the idea of fractals was first developed, it was not realized that they have any connection with chaos theory. During the 1970s, however, scientists noted that they are mathematical cousins. Both are concerned with irregularities but from slightly different points of view. Chaos is concerned with irregularities in dynamic systems, whereas fractals are concerned with geometrical irregularities. The existence of the often beautiful complex patterns of fractal theory leads us to the realization that there is no theoretical difficulty about the formation of complex structures in dynamic systems. Fractals such as snowflakes and cauliflowers have a considerable amount of order, but they can be formed in chaotic dynamical systems. The second law only imposes the condition that there must be an increase in disorder in the entire system, the ordered states being intermediates in the overall process.

It is often stated that two scientific advances, the concepts of relativity and of quantization, were the two greatest scientific achievements of the twentieth century. A good case can be made for regarding chaos theory as of equal significance for our understanding of the universe around us.

The idea of chaos theory was nicely summed up by Mark Twain long before it was thought of: "It is difficult to make predictions, especially about the future."

11

# SCIENTIFIC TRUTH

There are several different kinds of truth.[1] Sometimes artistic people use the word *truth* in a rather vague way, as when they talk about the truth of a painting, a poem, or a work of fiction; we can call that aesthetic truth and need say no more about it here. Other kinds of truth are religious truth, legal truth, historical truth, and scientific truth. It is unfortunate that the same word *truth* is used for these different kinds of truth, since they are all different ideas. Religious truth differs sharply from the other kinds of truth, since faith plays the primary role and evidence is of little importance.

In many other kinds of truth, the perceived facts are of dominant significance. In a court of law a witness is required to speak "the truth, the whole truth, and nothing but the truth." Witnesses are usually required to stick to the "facts," by which is meant the perceived facts. Witnesses to an event such as a traffic accident, even if they are intelligent, observant, and honest, often give inconsistent accounts of the facts as perceived by them. At the end of a trial a lawyer or judge may summarize the evidence and interpret it, and then a jury may reach a verdict. It is hoped that the verdict corresponds to the truth, and no doubt it sometimes does. Occasionally it emerges later that the verdict that was reached was not based on "the truth." Under ideal conditions in a well-conducted trial we may be reasonably confident that the verdict corresponds with the truth, but there can never be certainty. Criteria have been applied in the course of the trial, and it is hoped that no serious error has been made. As in all kinds of truth, we just have to do the best that we can.

With historical truth the situation is somewhat similar except that we are usually concerned more with documentary evidence and less with evidence from eyewitnesses. Again, we sift through the evidence and come to what we think is a reasonable conclusion.

There are two essentially different ways of approaching the truth. The method exclusively used in the early years of civilization was the *intuitive* method—the truth as we feel it instinctively. Early humans believed that Earth is flat and that the Sun goes around us. In the Old Testament, for example, this view is taken for granted, and it took many centuries for it to be corrected. The intuitive method was largely used by Aristotle, who still ranks as a scientist of great distinction, even though he sometimes went badly astray. He felt that certain things *must* be true and often did not bother to test them. As a result of his dependence on intuition, Aristotle was less successful in physics, where theory is always important, than in biology, which was then a descriptive science.

The intuitive method continued to be the most important method until the Renaissance, when more emphasis began to be placed on the *empirical* method. This is the method on which scientists, as well as many others, depend for their final conclusions. This method places great emphasis on observation and experiment as well as other kinds of evidence. The essential feature of the method is that it *works in practice*, which is the best method we can think of for our particular purposes.

Intuition is still important today, as long as it is used wisely. People use the word *intuition* in different ways, and we must be clear about just what we mean by it. The *Shorter Oxford Dictionary* gives as its first definition of intuition "immediate apprehension by the mind without reasoning." That is a good start, but I think we need a little further clarification and consideration. In some of Agatha Christie's detective stories, her detective Hercule Poirot makes some interesting comments about intuition. In *The ABC Murders*, for example, he remarks, "In a well-balanced mind there is no such thing as an intuition—an inspired guess," and later he says, "But what is often called an intuition is really based on logical deduction or experience. When an expert feels that there is something wrong about a picture or a piece of furniture . . . he is really basing that feeling on a host of small ideas and details." In other words, intuitive ideas spring to one's mind on the basis of our background knowledge and the previous thinking we have done. Poirot went too far in implying that a rational person never has an intuitive thought. The point is rather that such a person treats intuitive ideas in a critical but constructive way, bringing the light of reason to bear on them. The point can be made in another way by saying that there is no objection to using intuition in one's preliminary thinking, but that in coming to a final conclusion one should always use the empirical method.

It is thus an oversimplification to say that in science and in other scholarly fields, empirical methods have *entirely* taken the place of intuitive ones. On the contrary, scientists who have made important and original contributions always make good use of intuition, and some highly creative scientists are mainly intuitive in their approach to their research. Distinguished Dutch chemist Jacobus Henricus van't Hoff (whom we met earlier) was a great exponent of the use of imagination and intuition by scientists. When he delivered his inaugural lecture at the University of Amsterdam in 1878, he used as his title "Imagination in Science." He said that he had made a special study of the way in which the great scientific advances of the past had been made and stressed that it is important to use imagination as well as observation, real progress never being made by slavish examination and analysis of data. His peroration was a quotation from English historian Henry Thomas Buckle (1821–62): "There is a spiritual, a poetic, and for aught I know a spontaneous and uncaused element in the human mind, which ever and anon, suddenly and without warning, gives us a glimpse and a forecast of the future and urges us to see truth, as if by anticipation."[2] Van't Hoff always made certain, however, that in the last analysis his ideas were consistent with the experimental evidence.

Nonscientists often criticize scientific theories on the grounds that they seem to be unreasonable, by which they mean that they are counterintuitive. Einstein was criticized by philosophers on the grounds that his theory of relativity defied their intuitive ideas about space and time. Planck's quantum theory was criticized, even at first by some eminent scientists, on the grounds that the idea of packets of energy seemed unreasonable. The basic error that the critics were making was to use intuition as a valid way of testing a scientific theory. It is essential to realize that intuition may be of great help in formulating a theory *but is thoroughly unreliable for testing it.*

Reliance on *authority* is sometimes regarded as a path to the truth, but it is a derived path in which we rely on someone else to formulate our ideas. Religions often urge or require their adherents to accept without question the authority of religious writings and religious leaders. It must be admitted that in practice scientists also depend to a great extent on authority. The quantity of scientific information is now so vast, and the volume of theory so intricate and widespread, that no scientist today can hope to verify more than a tiny fraction of it. No one could possibly check much of what appears in the scientific literature, even within a narrow branch of science. We have to depend to a great extent on the authority of others, which means that we have to assume tentatively that what we read is correct—in other words, we have to have a certain amount of faith in others. Of course, we are always alert to the fact that we may be reading something that is mistaken or even fraudulent. But further research usually identifies errors or fraud.

Einstein's famous formula, $E = mc^2$, provides an interesting example of the temporary reliance that scientists sometimes place on intuition and authority. The equation appeared in 1905, but convincing experimental evidence for it only emerged in the early 1930s. During the intervening period some scientists accepted it tentatively, since the theory was consistent with the experimental evidence and had a valid logical structure. Moreover, it seemed intuitively to be reasonable, and Einstein appeared to many to be a reputable authority. At the same time, scientists always had in mind that the theory—indeed, any theory—might prove to be wrong. Some of the latest theoretical ideas, such as superstring theory, are in a similar situation; there is still not enough convincing evidence for them, and scientists treat them with particular caution.

There is no such thing as a scientific method that is used only by scientists. Instead there is a method used by scholars in all fields and also by medical practitioners, lawyers, engineers, policemen, and plumbers. For want of a better term I call this method the *judicial method*, or the *academic method*. Several important criteria are essential to this method. One is *objectivity*, according to which we should consider only the relevant evidence and put aside our personal feelings and prejudices. Another criterion essential to the judicial method is *coherence*. Any conclusion that is reached should be internally consistent; everything should hang together logically. In the field of law, this is particularly true of a complex judgment, which must be self-consistent. Otherwise the judgment simply would not *work*; it would be subject to widespread criticism and would lead to much confusion. Exactly the same should be true of a scientific publication, a medical report, or a government document. Thirdly, any approach to the truth must be recognized as *provisional*, and to some degree *approximate*. A scientific theory, a medical report, or a legal judgment may be the best that can be done in the circumstances, but it may well be wrong. It may be badly wrong, as in the case of people who are wrongly convicted of a crime or whose ailments are incorrectly diagnosed. We can approach the truth, but we must include a margin of error.

We can sum up by saying that the main criteria of the judicial method are that conclusions must be reached by careful consideration of all the evidence and that they are coherent, which means that they are logical and self-consistent. We should be prepared tentatively to use intuition or to rely on authority, as long as our final conclusions are consistent with the evidence and are coherent. It is here that those of us who apply the judicial method rigorously part company with many religious people, who rely on authority and intuition and fail to establish consistency with the evidence and the coherence of their beliefs.

▲▲▲

We should consider the judicial method in greater depth in relation to scientific investigation.[3] Scientific ideas have to be objective because they are *public* and open to the scrutiny of anyone interested. The convention in science is that a piece of research is not officially recognized until it has been published in a scientific journal. Normally a paper is not accepted for publication until it has been given some scrutiny by the editor of the journal or his delegate, who acts as a referee; this has come to be known as *peer review*. To get a scientific paper published we are required to include enough detail that readers are in a position to repeat the experiments and check the mathematics.

This convention of requiring science to be public is an important safeguard, since if anything is wrong with the research it is soon discovered. Coherence in science means that it must all hold together and form a self-consistent and logical network. Scientists in many diverse fields—physics, chemistry, astronomy, geology, and biology—have brought together, from a variety of independent sources, compelling evidence for the age of the universe, and for the ages of many of the structures within it, such as Earth and the living things on it. Confidence in the validity of the scientific method is further supported by the many technological advances that have been made possible by the advance of pure science.

Science is a complex network of concepts, models, and theories that are all closely intertwined. As a result, the whole structure is much stronger than any single component of it. There may be some weak structural components, but the rest of the structure is strong enough for all of it to hold together, just as a large building will not collapse if we remove a few struts. Eventually the weak components will be identified and replaced or repaired. A good example of repair is provided by the theory of relativity. It did not completely replace Newton's laws of motion, which are still serviceable for most purposes. Instead it adjusted and elaborated them so that they apply to the special case of objects moving at enormous speeds and deal with a few other problems outside the scope of Newton's equations.

As scientists in the various disciplines continue their studies and probe in more and more detail, it becomes increasingly clear that the same basic laws of nature are valid for all aspects of our universe.

The most basic of the sciences is physics, based solidly on mathematics and concerned with general relationships that apply to all matter. Originally, physics was mainly concerned with matter that is within easy reach. Galileo and Newton, for example, did much to clarify the motions of moving bodies. One of the most significant things that Newton did was to show that our solar system follows exactly the same laws that apply to terrestrial bodies. Later work has confirmed the conclusion that astronomy and cosmology are firmly based on the laws of physics and are essentially branches of physics.

About a century ago it became possible to probe much more deeply into the microscopic world of atoms and molecules. We were then able not so much to reject the old principles but to modify and clarify them. What again became increasingly clear was that the methods of physics applied more widely than had been supposed. The basic principles of mechanics, for example, suitably modified by the quantum theory, would explain the behavior of atoms and molecules. That meant that chemistry was also covered by the basic principles discovered by physicists and is thus, in a sense, another branch of physics. That does not mean that chemists should all become physicists or that physicists must take over chemistry. Because of the great complexity of nature, it is most convenient for chemistry to remain a science in itself, but chemists recognize that their science must follow the same basic rules as physics.

Geologists and biologists have also come to realize that their sciences must conform to the basic principles that have been established by the study of physics and chemistry. Geologists rely on chemistry to analyze the composition of many formations on Earth and on the Moon; they rely on physics to understand how rocks, mountains, and oceans were formed and continue to change. And they have worked with anthropologists to gather much of the evidence for evolution.

For many years biologists tended to think that the laws of nature that apply to inanimate objects did not necessarily apply to living systems, which were supposed to have some special characteristics. They pointed out, for example, that there is an increase in order accompanying the evolution of species, in apparent violation of the second law of thermodynamics. We now know, however, that this is a misinterpretation: the second law applies to the universe as a whole and allows increases to occur provided that they are accompanied by decreases in other parts of the universe. Similarly, other apparent violations of nature's laws have been explained as our understanding of scientific principles has matured.

In particular, some significant scientific discoveries found in the middle of the twentieth century made this conclusion inevitable. One of these was the elucidation, in 1953 by Perutz, of the structure of the protein hemoglobin by the x-ray diffraction method. Another was the suggestion, by Crick and Watson, also in 1953, of the double-helix structure of DNA and their explanation of how replication occurs. The structures of proteins and of DNA rely on the physical forces of amino acids and of nucleic acid bases to maintain their structures. These contributions played the crucial role in leading to the establishment of essentially a new branch of science, molecular biology. This field of study has finally brought physics, chemistry, and biology into complete harmony.

Still, it must always be borne in mind that scientific knowledge is approximate and provisional; anyone who has studied the history of science is very conscious of this. At various times scientists have decided that most of the problems of science had been solved. Even as great a scientist as Lord Kelvin fell into this error, suggesting toward the end of the nineteenth century that only a few trivial details of physical theory needed to be worked out before everything would be understood.[4] In the future, he thought, there would not be much for scientists to do besides making more accurate measurements. Kelvin never suspected that great developments would spring from the discovery of radioactivity and x-rays and that two comprehensive theories, of quanta and of relativity, would lead to fundamental changes in our scientific thinking and to remarkable scientific and technical advances.

The scientific events of the twentieth century have made us wary of making such predictions. In his interesting book *What Remains to Be Discovered*, Sir John Maddox has made predictions about science that are in striking contrast to those of Kelvin and that are overwhelmingly convincing.[5] Far from concluding that almost everything has been solved, he emphasizes that the number of questions to be answered increases rather than decreases as discoveries are made. To explain this apparent paradox Maddox uses the analogy of the unpacking of a nest of Russian dolls. When we unscrew the outermost one, we discover another, and when that is unscrewed another is revealed—and so it goes on. Unscrewing the early theory of the atom revealed electrons; further unscrewing discovered radioactivity, with its promise that atoms are not indivisible. Further unscrewing revealed neutrons and other subatomic particles. In a broadcast talk on October 1, 1939, Winston Churchill made a comment that might have been said about scientific knowledge: "It is a riddle wrapped in a mystery inside an enigma." In fact, it was the behavior of Russia that Churchill said was a riddle wrapped in a mystery inside an enigma![6]

▲▲▲

Is there any reason why we should not use the judicial method in dealing with problems other than scientific ones? In my book *Science and Sensibility* I argue in some detail that we should always use it in reaching all our final conclusions.[7] The judicial method is already used widely by scholars in fields other than science. They also do their work by carefully considering evidence in an unprejudiced manner and drawing logical conclusions. The same is true of those concerned with practical matters, such as doctors, nurses, car mechanics, plumbers, and electricians. All of them know that they will achieve the best results if they follow what I have called the judicial method.

By contrast, the intuitive method, which comes to us instinctively and was used exclusively in earlier times, is often unreliable. It is helpful in our preliminary thinking, and many scientists have made good use of it. When faced with arriving at a final conclusion, however, they find it essential to test their ideas by the judicial method, paying due regard to the observational and experimental evidence. Scientists and other scholars have also learned that although it is often helpful and convenient to pay attention to the conclusions of authorities, it is unwise to rely on them completely.

Although intuitive methods and complete reliance on authority are no longer relied upon by most scholars—and by most practical people—they are still widely used in other areas. In particular, politicians and senior government bureaucrats are still largely wedded to such methods. This is partly because they for the most part blindly adhere to an adversarial system, wherein they tend to join a particular political party and then accept the authority of its leaders. It would be much better if they would put their prejudices aside—particularly their allegiance to particular races, countries, and political parties—and would select evidence and arrive at their decisions in an unbiased way. The opinions of political parties are never based on all of the evidence, but rather on selected evidence. In *Science and Sensibility* I have given many examples of how one can arrive at a completely false conclusion by selecting the particular evidence that supports one's case. As eminent historian Lord Macaulay put it, "He who is deficient in the art of selection may, by showing nothing but the truth, produce all the effects of the greatest falsehood."

Science has had remarkable success in arriving at the concept of a harmonious universe by the use of the judicial method of investigation. I am convinced that the world would be a much better and happier place if all of us followed the same method.

# NOTES

In these notes I have not included many references to original scientific papers, which only specialists will want to look at. A number of such references are collected in two of my previous books: *The World of Physical Chemistry* (Oxford: Oxford University Press, 1993) and *To Light Such a Candle* (Oxford: Oxford University Press, 1998).

## CHAPTER 1

1. Benjamin Franklin, "Of the Stilling of Waves by Means of Oil," *Philosophical Transactions of the Royal Society* 64 (1774): 445–60. This work, as well as other works by Franklin, is described in a most interesting way by Charles Tanford, *Ben Franklin Stilled the Waves: An Informal History of Pouring Oil on Water with Reflections on the Ups and Downs of Scientific Life in General* (Durham, NC, and London: Duke University Press, 1989). For even more details about the oil experiments of Franklin and others, see C. H. Giles, "Franklin's Teaspoonful of Oil," part 1 of "Studies in the Early History of Surface Chemistry," *Chemistry and Industry*, November 8, 1969, pp. 1616–24; C. H. Giles and S. D. Forrester, "Wave Damping: The Scottish Contribution," part 2 of "Studies in the Early History of Surface Chemistry"), *Chemistry and Industry*, January 17, 1970, pp. 80–87.

2. The calculation that substantiates this claim is as follows:

1 liter of water = 55.5 moles = $55.5 \times 6.022 \times 10^{23}$ molecules = $3.3 \times 10^{25}$ molecules.

In a chain of water molecules, held together by hydrogen bonds, the distance between neighboring oxygen atoms is about 300 pm ($3 \times 10^{-10}$ m). This is the length of each link in the chain, the total length of which is therefore

$$3.3 \times 10^{25} \times 3 \times 10^{-10}\text{ m} = 10^{16}\text{ m} = 10^{13}\text{ km}.$$

(This is a little more than a light-year, which is $9.462 \times 10^{12}$ km).

The distance to the Moon is about $4 \times 10^5$ km (about 1.3 light-seconds), so that the distance there and back is $8 \times 10^5$ km. The chain of $H_2O$ molecules would thus go to the Moon and back $10^{13}/(8 \times 10^5) = 1.25 \times 10^7$ times, or 12.5 million times.

3. This stipulation proved impractical; as a result, later Nobel Prizes were sometimes awarded even for work done decades earlier.

4. G. N. Lewis, *Valence and the Structure of Atoms and Molecules* (New York: Chemical Catalog, 1923).

## Chapter 2

1. For an excellent new annotated translation of Newton's *Principia* see *The Principia: A New Translation,* trans. I. Bernard Cohen and Anne Whitman (Berkeley and Los Angeles: University of California Press, 1999).

2. Isaac Newton, *Opticks: A Treatise on the Reflections, Refractions, Inflections & Colours of Light* (London: Innys, 1704). Unlike the *Principia,* which was in Latin, this work was written in English. For a facsimile reprint, see Isaac Newton, *Opticks: or Treatise on the Reflections, Refractions, Inflections & Colours of Light,* based on the 4th ed. (New York: Dover, 1952), with later additional reprints. This reprint has a foreword by Albert Einstein, an introduction by Sir Edmund Whittaker, a preface by I. Bernard Cohen, and an analytical table of contents prepared by Duane H. D. Roller.

3. Benjamin Franklin, *Experiments and Observations on Electricity, Made at Philadelphia in America* (London: E. Cave, 1751). Tom Tucker, *Bolt of Fate: Benjamin Franklin and His Electric Kite Hoax* (New York: Public Affairs, 2003), makes the claim that Franklin did not actually carry out the kite experiment.

4. Alexander Volta, "On the Electricity Excited by the Mere Contact of Conducting Substances of Different Kinds: In a Letter from Mr. Alexander Volta, in the University of Pavia, to the Rt. Hon. Joseph Banks, Bart., K. B. P. R. S." *Philosophical Transactions of the Royal Society* 92 (1800): 403–31.

5. Nicholson, J., "Account of the New Electrical or Galvanic Apparatus of Sig. Alex. Volta, and Experiments Performed with the Same," *Journal of Natural Philosophy, Chemistry, and the Arts* (Nicholson's Journal) 4 (1800): 179–87. Appearing in July 1800, this paper was published in fact shortly before the publication of the paper by Volta.

6. J. Clerk Maxwell, *A Treatise on Electricity and Magnetism*, 2 vols. (Oxford: Oxford University Press, 1873), was the culmination of a long series of papers by Maxwell on electromagnetic theory, the first appearing in 1855.

7. Chapter 5 of Keith Laidler, *To Light Such a Candle* (Oxford: Oxford University Press, 1998), pp. 159–200, traces in some detail the practical consequences of Maxwell's theory of electromagnetism. In particular, his theory soon led to the invention of radio.

## Chapter 3

Energy in all its aspects is considered in more detail in Keith Laidler, *Energy and the Unexpected* (Oxford: Oxford University Press, 2003).

1. In ordinary speech we often use the word *weight* when we are thinking about the *mass* of an object; however, these are different concepts. The weight of an object is the force due to the gravitational field; thus it is less on the Moon than on Earth because the lighter Moon exerts a weaker gravitational force on it. Conversely, its weight is much greater on a much larger planet such as Jupiter. The mass of an object is the amount of material contained in it and is therefore the same on the Moon, on Earth, and on Jupiter.

2. Carnot, Sadi, *Réflexions sur la puissance motrice du feu et sur les machines propre à développer cette puissance* [Reflections on the Motive Power of Fire and on Machines Designed to Develop That Power] (Paris: Bachelier, 1824).

3. Ibid.

4. For more on Maxwell's demon see Henry S. Leff and Andrew F. Rex, *Maxwell's Demon* (Princeton, NJ: Princeton University Press, 1990); Hans

Christian von Baeyer, *Maxwell's Demon: Why Warmth Disperses and Time Passes* (New York: Random House, 1998); paperback edition, *Warmth Disperses and Time Passes: The History of Heat* (New York: Modern Library, 1999).

## CHAPTER 4

1. Many of these are collected in Walter Gratzer, *Eurekas and Euphorias: The Oxford Book of Scientific Anecdotes* (Oxford: Oxford University Press, 2002), pp. 67–71.

2. It has been said of Van de Graaff that his name holds the record among scientists as being most often misspelled. (I think that I have it right!)

3. Michael Frayn's play *Copenhagen* (Colchester, Essex: Methuen Drama, 1998), a recent stage success, makes some reference to the Copenhagen interpretation but is primarily a discussion between Bohr and Heisenberg, along with Mrs. Bohr, about Heisenberg's role in the German atomic bomb program.

4. For other examples of Schrödinger's eccentric behavior see Gratzer, *Eurekas and Euphorias*, pp. 191–94.

## CHAPTER 5

1. According to the theory of relativity, of course, it is in reality impossible to attain the speed of light, because the mass traveling at that speed would become infinite and the kinetic energy would be infinite. The calculation in the text does not take this factor into account; it just relates to the conversion of the normal mass of a person moving at the speed of light.

2. The discrepancy between the two times is due to Earth's rotation.

3. The circumstances regarding these observations are not as straightforward as they seem from this brief account; for further details with more insight see Walter Gratzer, *Eurekas and Euphorias: The Oxford Book of Scientific Anecdotes* (Oxford: Oxford University Press, 2002), pp. 121–23.

4. Other stories about Einstein are to be found in Gratzer, *Eurekas and Euphorias*, especially pp. 262–63.

5. A similar and even better story is told in Gratzer, *Eurekas and Eupho-*

*rias*, p. 53, about mathematician Norbert Wiener. Unable to find his house he invited the help of a small girl to whom he introduced himself. "Yes, Daddy," she said, "I'll take you home."

6. Other difficulties over relativity with Oxford philosophers, involving physics professor F. A. Lindemann, are recounted in Gratzer, *Eurekas and Euphorias*, pp. 93–97.

## Chapter 6

Great help in understanding the complicated story of nuclear particles is provided by John Gribbin, *Q Is For Quantum: An Encyclopedia of Particle Physics* (New York: Free Press, 1998).

1. A slight qualification is needed here, since sometimes a particle is its own antiparticle. A neutron, for example, having no charge, is its own antiparticle.

2. CERN is an acronym for the organization's former name, Conseil Européen pour la recherche nucléaire.

3. The mass unit used here, MeV/$c^2$, which relates to the equation $E = mc^2$, is the one often used by particle physicists. It represents the energy (MeV) divided by the square of the speed of light, $c^2$. One kilogram is equal to $5.699 \times 10^{29}$ MeV/$c^2$. Remember that 1 MeV = $10^6$ eV = 1,000 eV and 1 GeV = $10^9$ eV = 1,000 MeV.

## Chapter 7

1. The idea of continental drift is developed in a most interesting way in a popular book by Simon Winchester, *Krakatoa: The Day the Earth Exploded, August 27, 1883* (New York: HarperCollins, 2003). An excellent account is also given in Peter D. Ward and Donald Brownlee, *Rare Earth: Why Complex Life Is Uncommon in the Universe* (New York: Copernicus, 2000), esp. chap. 9, "The Surprising Importance of Plate Tectonics," pp. 191–220.

2. What is true, however, is that similar experiments were later made at Newton's suggestion in St. Paul's Cathedral in London. Newton described these experiments in his *Principia Mathematica*; see *The Principia: A New Translation*, trans. I. Bernard Cohen and Anne Whitman (Berkeley and Los Angeles: University of California Press, 1999), pp. 756–61.

3. In 2003, spaceships travel at the same average speed as they did forty years previously. They have usually reached the Moon, about 384,000 km (238,600 mi) from Earth, in about 76 hours, which means that their average speed was about 5,100 km $h^{-1}$ (3,200 mph). When Earth and Mars are closest together the distance between them is about $5 \times 10^7$ km (roughly $3.1 \times 10^7$ mi). Travel to Mars with unmanned craft has taken at least 5 months, or 3,600 hours, which means an average speed of 13,900 km $h^{-1}$ (8,640 mph), substantially greater than for flight to the Moon. This is still a good deal less than the speed of 36,000 km $h^{-1}$ (22,400 mph) suggested for the hypothetical spaceship of the future, a speed that would get us to the Moon in a little more than 10 hours. NASA has plans to use rocket-powered spacecraft that might have average speeds of about that speed, which would enable us to reach Mars in about 1,390 hours, or less than 2 months. It is expected that the development and testing of such rockets might take 10–20 years.

4. If the universe is 12 billion years old, the galaxy has therefore traveled 10 billion light-years in 2 billion years. Has it gone five times as fast as the speed of light? The answer, provided by Einstein's special theory of relativity, is in outline as follows. The galaxy cannot travel faster than the speed of light but must have traveled at nearly the speed of light. Under these conditions the time frame is altered, in the sense that a fast-running clock runs slowly and can travel more than one light-year for every year that it records. According to the equation for Einstein's special theory of relativity, a clock on a galaxy traveling at 98 percent of the speed of light will travel five light-years but only record that it has traveled for one year. From an external observer it therefore appears to have been traveling five times as fast as the speed of light.

5. Some people, such as astronomer Sir Fred Hoyle, have made an issue of the fact that Jocelyn Bell did not share the prize, but Jocelyn Bell Burnell (as she is now known) generously takes no such position.

## CHAPTER 8

1. Here are a few more details about the value of the Hubble constant (mathematically represented $H$). When measurements were made from observatories on Earth, even at different positions in its orbit, the values obtained for the constant were rather widespread, although all pointed to an age of the universe in the billions of years. Recent measurements made

by astronomical satellites, particularly the European Space Agency satellite *Hipparcos*, provide more reliable values. Recent determinations range from about fifty to seventy-five kilometers per second per megaparsec (which can be written as km $s^{-1}$ $Mparsec^{-1}$).

This megaparsec unit, usually used by astronomers for the Hubble constant, requires explanation, and we need the following data:

$$\begin{aligned}
&\text{1 parsec} &&= 3.262 \text{ light-years} \quad \text{1 Mparsec} = 3.262 \times 10^{6} \text{ light-years}\\
&\text{Speed of light} &&= 2.998 \times 10^{8} \text{ m s}^{-1} \quad \text{1 year (y)} = 3.156 \times 10^{7} \text{ s}\\
&\text{1 light-year} &&= 2.998 \times 10^{8} \text{ m s}^{-1} \times 3.156 \times 10^{7} \text{ s}\\
& &&= 9.462 \times 10^{15} \text{ m} = 9.462 \times 10^{12} \text{ km}\\
&\text{1 Mparsec} &&= 3.262 \times 10^{6} \times 9.462 \times 10^{12} \text{ km}\\
& &&= 3.087 \times 10^{19} \text{ km.}
\end{aligned}$$

A Hubble constant of 1 km $s^{-1}$ $Mparsec^{-1}$ is thus

$$\begin{aligned}
(1/3.087 \times 10^{19})\ \text{s}^{-1} &= 3.239 \times 10^{-20}\ \text{s}^{-1} = 3.239 \times 10^{-20} \times 3.156 \times 10^{7}\ \text{y}^{-1}\\
&= 1.022 \times 10^{-12}\ \text{y}^{-1}.
\end{aligned}$$

The two extreme values quoted above therefore correspond to

$$\begin{aligned}
50 \text{ km s}^{-1} \text{ Mparsec}^{-1} &= 50 \times 1.022 \times 10^{-12}\ \text{y}^{-1} = 5.11 \times 10^{-11}\ \text{y}^{-1}\\
75 \text{ km s}^{-1} \text{ Mparsec}^{-1} &= 75 \times 1.022 \times 10^{-12}\ \text{y}^{-1} = 7.67 \times 10^{-11}\ \text{y}^{-1}.
\end{aligned}$$

If we take the second value and assume the Hubble constant to have remained the same since the big bang, the time that has elapsed since the galaxies started to move apart is

$$1/7.67 \times 10^{-11} \text{ years} = 1.30 \times 10^{10} \text{ years} = 13.0 \text{ billion years.}$$

The first, lower, value of $H$ leads to 19.6 billion years. The following table relates Hubble constants to ages:

| $H$ (km $s^{-1}$ $Mparsec^{-1}$) | Age of universe (in billions of years) |
|---|---|
| 50 | 19.6 |
| 60 | 16.3 |
| 70 | 14.0 |
| 75 | 13.0 |
| 80 | 12.2 |
| 100 | 9.8 |

The most recent work favors the lower values of the Hubble constant, which would suggest an age of 18 to 20 billion years. However, because of the gravitational attractions between the galaxies, the constant must have been bigger in the past than it is today. The age of the universe is therefore smaller than that obtained from modern measurements of the constant. The best theoretical treatments of the correction required from this gravitational effect suggest that the calculated ages should be reduced by 30 percent. Thus, instead of an age of 19.6 billion, many experts agree that a more realistic value is about 14 billion years.

The value of 12 billion years, which I have used in the main text, is therefore a conservative estimate. For further details see Martin Gorst, *Measuring Eternity: The Search for the Beginning of Time* (New York: Broadway Books, 2001).

2. See, for example, the books by Peter Atkins (*Creation Revisited*), John Gribbin (*In Search of the Big Bang*), Joseph Silk, and Steven Weinberg, listed under "Suggested Reading."

3. In fact, Darwin rather overestimated the age, which is now considered to be more like 135 million years. This caused him some embarrassment, as is well discussed in Gorst, *Measuring Eternity*, pp. 165–70, 303.

4. Cecilia Payne married Sergei Gaposchkin, a boisterous Russian astronomer, whom she had met in Europe when he was down on his luck. She managed to get him a post at Harvard, where he was no more than her assistant, although he never realized it; he was once heard to say, "Cecilia is an even greater scientist than I am." Some further anecdotes about her life are to be found in Walter Gratzer, *Eurekas and Euphorias: The Oxford Book of Scientific Anecdotes* (Oxford: Oxford University Press, 2002), pp. 243–45. One extra detail about her is that while still at school she tested the efficacy of prayer by a controlled test; she divided her examinations into two groups and prayed for success in one group only. Since she actually got better marks in the prayerless group, she became, and remained, a confirmed agnostic.

## CHAPTER 9

1. Mendel was born Johann Mendel, in northern Moravia, now a part of the Czech Republic. He adopted the name Gregor upon his ordination in 1847.

2. Richard Dawkins, *The Blind Watchmaker* (London: Longmans, 1986; London: Penguin, 1988; reprinted with an appendix, 1991), p. 116.

3. The numbers of base pairs in the chromosomes of different species are difficult to rationalize. On the whole, simpler species have fewer base pairs: in a small virus, for example, the number is about five thousand. The

simple newt, on the other hand, has 20 billion base pairs, many more than the 3 billion that we have.

4. Since the conditions for the evolution of higher forms of life are so stringent, it is possible that we humans, living on this Earth, are unique in the universe. This is argued very persuasively by Peter D. Ward and Donald Brownlee, *Rare Earth: Why Complex Life Is Uncommon in the Universe* (New York: Copernicus, 2000).

5. Paul Davies, *The Fifth Miracle: The Search for the Origin and Meaning of Life* (New York: Simon & Schuster, 1999).

## CHAPTER 10

1. There is an interesting story behind this mathematical contest. Poincaré's winning article was printed in 1889 in an issue of the *Acta Mathematica,* an increasingly influential journal. Soon after the issue appeared, one of its readers pointed out a significant error in Poincaré's treatment. This gave rise to serious complaints that another contestant, who had submitted a satisfactory treatment of the problem but one not as thorough as that of Poincaré, should have been the winner. The editor of the *Acta* and chief organizer of the contest, mathematician Gösta Mittag-Leffler (1846–1927), was naturally disturbed by this controversy, feeling that it would sully the reputations of those who had been concerned with selecting the winner of the contest—after all, why had they not detected the error in Poincaré's submission? Mittag-Leffler took rather drastic action: instead of publishing a correction in a subsequent issue he diligently tracked down to their destinations as many of the offending issues as he could, later destroying them. He was so successful at this that today only one of the issues is known to exist, and it is kept inaccessible under lock and key in the Mittag-Leffler Institute near Stockholm.

At the same time, Mittag-Leffler persuaded Poincaré to correct his error and to submit a revised paper that would be published in a subsequent issue of the *Acta* as if it had been the prize-winning paper. Poincaré never denied that he had made an error, although in view of the secrecy involved, no one knows just what it was. It presumably was not of vital significance, as Poincaré soon produced a satisfactory proof. Between the award ceremony and the appearance of the revised paper in 1890 there was a good deal of behind-the-scenes bickering, but it soon died down, and in the end Mittag-Leffler's cover-up had achieved its purpose.

# Chapter 11

1. For a general discussion of truth, see Felipe Fernández-Armesto, *Truth: A History and Guide for the Perplexed* (London: Bantam Press, 1997; London: Black Swan paperback, 1998).

2. Buckle wrote a monumental, and in its time highly influential, *History of Civilization in England,* which appeared from 1857 to 1861. In it he urged what he called a scientific method of writing history, taking into account a country's climate and other factors.

3. Scientific truth is further discussed in Roger G. Newton, *The Truth of Science: Physical Theories and Reality* (Cambridge, MA: Harvard University Press, 1997).

4. Lord Kelvin's opinion that by the end of the nineteenth century most scientific problems had been solved was given in a lecture at the Royal Institution in London on April 27, 1900, and published as "Nineteenth Century Clouds over the Dynamical Theory of Heat and Light," *Philosophical Magazine,* 6th ser., 2 (1901): 1–40. Similar opinions appeared in his book *The Baltimore Lectures on Molecular Dynamics and the Wave Theory of Light* (Cambridge: Cambridge University Press, 1904). The lectures referred to in the title were given at Johns Hopkins University in 1884. This book was reprinted in *Kelvin's Baltimore Lectures and Modern Theoretical Physics: Historical and Philosophical Perspectives,* ed. R. Kargon and P. Achinstein (Cambridge, Massachusetts: MIT Press, 1987). This volume also contains several essays on Kelvin's work in relation to modern physics.

5. John Maddox, *What Remains to Be Discovered: Mapping the Secrets of the Universe, the Origins of Life, and the Future of the Human Race* (New York: Simon and Schuster, Free Press, 1998).

6. The complete quotation from Winston Churchill's speech is, "I cannot forecast to you the action of Russia. It is a riddle wrapped in a mystery inside an enigma."

7. Keith J. Laidler, *Science and Sensibility: The Elegant Logic of the Universe* (Amherst, NY: Prometheus Books, 2004).

# GLOSSARY

**Absolute temperature.** More formally known as thermodynamic temperature, it is equal in practice to the Celsius temperature plus 273.15 K. Thus if the Celsius temperature is 25.00 degrees, the absolute temperature is 298.15 K, where K stands for (Lord) Kelvin, the originator of the absolute scale of temperature. The temperature zero Kelvin (known as *absolute zero*) is the lowest possible temperature, and it cannot quite be reached; temperatures of a tiny fraction of a degree over absolute zero have been attained.

**Academic method.** See **judicial method**.

**Adaptation.** In biology, the process in which an organism becomes adjusted to its environment.

**Allele.** Different forms of a given gene. It is often said that different humans have different genes, but this is not quite correct. All humans of the same sex have the same genes; the differences are really in the alleles.

**Amino acids.** Naturally occurring organic molecules containing nitrogen atoms as well as carbon and hydrogen atoms. Twenty amino acids constitute the building blocks for protein molecules.

**Amplitude.** When a vibration is occurring about an equilibrium position, the distance between the object at the two extremities of its motion is called the amplitude of the vibration. The term also applies to wave motion.

**Antiparticle.** According to modern field theory, for every fundamental particle there exists an antiparticle having the same mass but the opposite charge or chargelike property. For example, the electron and the positron (the positively charged particle of the same mass) are particle and antiparticle. By convention the commoner particle is called the particle, the other the antiparticle. Some neutral particles, such as the photon, are their own antiparticles.

**Atom.** The smallest part of a chemical element that still retains the basic characteristics of the element. An atom consists of a minute positively charged nucleus surrounded by a cloud of electrons, the number of electrons being equal to the number of protons (positively charged) in the nucleus.

**Atomic mass unit (AMU).** The unit for expressing the mass of an atomic nucleus, relative to the mass of exactly 12 for the isotope $^{12}C$ (carbon-12) of the carbon atom.

**Bacteria (*sg.*, bacterium).** Single cells with no nucleus. Each cell is surrounded by a cell wall. Some bacteria are agents of diseases such as cholera; others play a benign role, for example in recycling in which the decomposition of organic wastes occurs. Bacteria are routinely used in genetic engineering.

**Bacteriophage (phage).** Viruses that infect and destroy bacteria.

**Base.** (1) In chemistry, a substance capable of neutralizing an acid. (2) In microbiology, one of the building blocks of DNA and RNA. These building blocks consist of a nitrogen-containing unit, which combines with sugar and phosphate molecules to form a nucleotide. The four bases in DNA are adenine (A), guanine (G), cytosine (C), and thymine (T).

**Base pair.** The pair of bases formed when two nucleotides are held together by a weak bond (called a hydrogen bond) between complementary bases. In a DNA molecule, A is paired with T, and C with G.

**Big bang.** The event, concluded to have occurred at least 12 billion years ago, in which the universe was born by a rapid expansion from enormously dense matter. Today this explanation of the beginning of the universe is accepted by almost all astronomers and cosmologists. Since its original proposal by the Abbé Georges Lemaître, and its development by George Gamow and others, the theory has been supported by a vast amount of astronomical evidence.

**Billion.** A thousand million ($10^9$). The word was formerly used in the United Kingdom to mean a million million, but the thousand-million definition is now universally accepted. The prefix *giga-* means a billion (e.g., *gigabyte*).

**Bioengineering.** The application of the methods of engineering to biological problems. A common form of bioengineering is genetic engineering, and the two expressions are often used synonymously.

**Biotechnology.** The application of bioengineering to the production of substances of nutritional or medical importance, as well as to medical procedures of any kind.

**Black body.** A hypothetical substance that would absorb all frequencies of radiation falling on it; since it reflects no radiation it would appear black. No such substance actually exists, but a close approximation to it is a cavity in a metal with a small hole leading to it. Any radiation passing into the cavity has little chance of being reflected out. *Black-body radiation* is the radiation emitted by a completely black body; such radiation depends only on the temperature of the body.

**Black hole.** A celestial body of such an enormous mass that even photons (particles of electromagnetic radiation) are unable to escape from it because of gravitational attraction. It is distinguished from an ordinary star by the lack of radiation from it. Sufficiently massive stars eventually become black holes.

**Boyle's law.** According to this law, for a given amount of a gas at constant temperature, the product of the pressure and the volume is constant:

$$\text{pressure} \times \text{volume} = \text{constant}.$$

Thus if we double the pressure on a gas, the volume decreases by one-half. This law is obeyed approximately by all gases under certain conditions. A hypothetical gas that obeys the law is known as an ideal gas.

**Brown dwarf.** A small star; by conventional definition it is a star having less than 8 percent the mass of the Sun. Such stars are too small to allow ordinary nuclear reactions to occur within them, so they emit little radiation. A few have been detected by their emission of red radiation, but there may be many more that are as yet undetected.

**Cell.** The basic structural and functional unit of living systems. A cell is the smallest possible unit of life (not counting viruses, which are generally considered nonliving). All cells consist of a mass of cytoplasm surrounded by a cell membrane. Some cells have nuclei; bacterial cells do not.

**Celsius temperature.** This temperature scale is very similar (to within a hundredth of a degree) to centigrade temperature. In the old centigrade scale, now replaced by the Celsius scale, zero is defined as the temperature at which ice freezes at one atmosphere pressure, and one hundred degrees is the temperature at which it boils at the same pressure. The Celsius scale, designed to be very close to the centigrade scale, is defined with reference to two fixed values: the lower is absolute zero, and the upper is the triple point of water, which is the temperature at which ice, liquid water, and steam are at equilibrium. This definition has the advantage of not involving standard atmospheric pressure, which is an arbitrary quantity.

**Chaos.** (1) The complete chaos that will exist at the end of time, when, according to the second law of thermodynamics, the universe will be at its state of maximum entropy or maximum probability. (2) A more limited form of chaos exists with certain processes, the outcome of which, as explained by chaos theory (see chap. 11), is impossible to predict.

**Charles's law.** See **Gay-Lussac's law**.

**Chromosomes.** Threadlike bodies that carry a living being's genes; they are present in the nuclei of cells. If unraveled into a straight line, they have an average length of about 5 cm (2 in). They carry the genetic information of the cell in their DNA, which is organized into linear arrangements of genes. The location of genes on particular chromosomes and their relative positions along the chromosome have been mapped by breeding experiments and more recently by techniques for probing specific DNA sequences. Each species has a particular number of chromosomes: humans have twenty-two pairs plus two additional chromosomes, X and Y, that are sex-related. A female has two X chromosomes, a male has an X and a Y.

**Clone.** A population of genetically identical organisms or cells derived from an original single organism or cell. The word is also applied to populations of DNA molecules or viruses copied from a single parent.

**Codon.** A triplet of nucleotides that is part of the genetic code and leads to the synthesis of a particular amino acid that adds on to a growing chain to become a protein molecule.

**Coherence.** In general, internal consistency. In physics, used with reference to electromagnetic waves; these are said to be coherent when they are in step with one another, in the sense that their troughs and crests all coincide in space and time.

**"Cold fusion."** In 1989 it was announced that a certain electrochemical technique would cause large amounts of heat to be produced with the expenditure of small amounts of electrical energy. The process involved the electrolysis of a heavy water solution with the use of palladium electrodes. Within a few years conclusive experiments by many other scientists showed that the original claim could not be sustained.

**Continental drift.** The movement of large landmasses with respect to one another over long periods of time. The original idea has been developed in terms of the modern theory of plate tectonics.

**Cosmic microwave background radiation.** Thermal black-body radiation that pervades the universe and is a relic of the big bang, corresponding to a temperature of about 2.7 K.

**Cosmology.** The scientific study of the structure and function of the universe, including its origin and evolution. Some religions claim to have cosmologies, but all cosmologies based on religious writings have been proved beyond reasonable doubt to be wrong.

**Dark matter.** Matter in the universe that is inferred to exist because of its gravitational attraction rather than by direct observation. About 90 percent of each galaxy, including our own, seems to be composed of dark matter. Neutrinos and brown dwarfs may be constituents of dark matter.

**Darwinism.** The theory of evolution that was put forward by Charles Darwin in 1859. The essence of the theory is that evolution comes about as a result of natural selection, a process in which environmental factors favor the survival of species that possess certain advantageous features. All subsequent work has shown that this idea is essentially correct. The modern version of Darwin's theory, with evolution interpreted in terms of genes, is sometimes referred to as neo-Darwinism.

**Data (*sg.*, datum).** Facts that have been obtained by observation or experiment and are therefore "given." *Datum* is a Latin word meaning "a thing given."

**Deuterium.** The isotope of hydrogen in which the nucleus of the atom consists of a proton and a neutron (instead of just a proton, as in ordinary hydrogen). It is given the symbol D or ${}^{2}_{1}H$. Heavy water, $D_2O$, is a form of water; in its molecule an oxygen atom is combined with two deuterium atoms. The chemical properties of deuterium and its compounds are much the same as those of the compounds of ordinary hydrogen.

**Disintegration (radioactive).** A process in which an atomic nucleus emits high-energy particles ($\alpha$- and $\beta$-particles) and electromagnetic radiation ($\gamma$-radiation), forming another nucleus.

**Dissipation of energy.** The wastage of energy that according to the second law of thermodynamics must occur when any spontaneous process takes place. The energy is not lost (in accordance with the first law) but is converted into forms in which it cannot do useful work (e.g., low-temperature heat).

**DNA (deoxyribonucleic acid).** The genetic material of nearly all living systems. The term applies to a class of large molecules composed of two complementary chains of nucleotides normally wound in a helix. DNA molecules, by controlling the synthesis of protein molecules, are the carriers of genetic information.

**Doppler effect.** The shifting of the frequency of a wave resulting from the relative motion of the observer and the source of the wave. It is observed with both sound and light waves. The Doppler effect is a key tool is measuring the speeds of astronomical objects (the redshift).

**Double helix.** The two-stranded, coiled structure of DNA. Each chromosome consists of a double helix, portions of which constitute the individual genes.

**Electromagnetic radiation.** Radiation emitted in the form of an electromagnetic wave, as opposed to particle radiation, in which there is a stream of particles having mass. Electromagnetic radiation can be regarded as made up of a stream of particles, the photons, which have zero mass (when they are at rest, which they never are).

**Electromagnetic theory.** The theory, originally proposed by James Clerk Maxwell, that treats light and other forms of radiation in terms of electric and magnetic fields. These are the fields of force that exist in the neighborhood of electric currents and magnets. Electromagnetic field theory is now so well established that it can be regarded as a cosmological law.

**Electron.** One of the fundamental building blocks of nature. We arbitrarily regard its charge as negative and say that it has a single negative charge—that is, that its charge is –1. This charge is equal to that on the proton but of opposite sign. The mass is about one two-thousandth of that of the proton.

**Electron volt.** The energy acquired by an electron when it passes through a voltage drop of one volt. One electron volt is roughly the energy that is carried by one photon of visible light.

**Elementary particle.** Also called a fundamental particle, a particle that cannot be broken down into smaller particles. Just which particles are taken to be elementary depends on how far science has progressed and on the amount of detail we want to go into. For many purposes of this book it is satisfactory to consider electrons, protons, and neutrons to be elementary particles. Recent work, outlined in chapter 6, has revealed that there are smaller particles.

**Empirical equation.** Also known as an empirical law, a mathematical equation that expresses a relationship between experimental data; Boyle's law is an example.

**Empirical law.** See **empirical equation**.

**Energy.** A physical property that relates to the capacity of a system to perform mechanical work. Energy occurs in various forms, such as kinetic energy (concerned with motion) and potential energy (concerned with position). The law of conservation of energy (the first law of thermodynamics) states that the total energy in the universe remains constant. In applying this principle one must bear in mind that there is an interconversion of mass and energy as well as the fact that in nuclear processes this effect is significant.

**Entropy.** A measure of the degree of disorder and inherent probability in a given system. The reason that a process occurs naturally is that the total entropy (of the system plus its surroundings) increases, which means that the final state is more probable than the initial state. These ideas are combined in the second law of thermodynamics, according to which the entropy of the universe increases as time goes on.

**Evolution.** The process by which organisms have developed and are developing from earlier and more primitive forms. Evidence for evolution comes from studies of fossil records, from observations of living things, and from many studies in biochemistry and physiology. Today evolution is still occurring in a very evident way: in the adaptation of bacteria to new drugs. The central theme of evolutionary theory is still the theory proposed in 1859 by Charles Darwin. In the late twentieth century there have been great advances in the study of classical and molecular genetics, supplementing Darwin's ideas and leading to the new theory of neo-Darwinism. The theory of evolution is now a major unifying concept in most areas of biology.

**Feedback.** The output of a process fed back to another process. For example, if a process is occurring in stages and there is positive feedback, some of the output is added to the input, with the result that there is instability. With negative feedback, some of the output is subtracted from the input, causing the system to settle down to a stable state.

**Field.** A region in which any kind of physical force is exerted is called a field, which is a physical property extending through space. For example, a magnet is affected by a current flowing through a neighboring wire, and there is said to be a magnetic field around the wire. There is also an electric field around a wire carrying a current, and we refer to the two fields as an electromagnetic field. Similarly, there are both electric and magnetic fields (i.e., an electromagnetic field) around a magnet.

**First law of thermodynamics.** This law states that energy can be neither created nor destroyed; in other words, the energy of the universe remains constant.

**Fundamental particle.** See **elementary particle**.

**Galaxy.** A group of an enormous number of stars, typically about $10^{11}$.

**Gamete.** An element existing in a sex cell, able to unite with another in sexual reproduction.

**Gamma ($\gamma$-) rays.** Electromagnetic radiation of enormously high frequency and therefore of high energy. Gamma rays are often emitted when nuclear processes occur and can also be generated by other high-energy processes.

**Gay-Lussac's law.** Often called Charles's law, this law states that at a constant pressure the volume of a gas is proportional to the absolute temperature.

**Gene.** The fundamental unit of inheritance and function. At the molecular level, a gene is made up of a sequence of nucleotides on a particular site on a chromosome. It is difficult to define a gene more precisely than to say that it is a specific section of a DNA molecule that codes for a recognizable unit that plays a role in inheritance and metabolism. Genes direct the formation of the proteins that a cell uses to function, repair itself, and divide. The number of base pairs in a gene varies widely. Typically a gene has 20,000 to 50,000 base pairs, but one, related to the operation of the human brain, is reported to have 2.3 million base pairs; when written out in terms of letters A, G, T, and C it stretches the length of a long corridor. Many human disorders are caused by flaws in genes or the absence of genes.

**Genetic code.** The code that relates the codons (triplets of nucleotides) to the specific amino acids.

**Genetics.** The study of heredity and the variation of inherited characteristics. Today genetic studies proceed along six main pathways: (1) *experimental breeding*; (2) *pedigree analysis*, in which specific traits are traced in a family line; (3) *cytogenetics*, the study of the chromosomes and other cellular structures; (4) *biochemical genetics*, the study of molecular processes relating to inheritance; (5) *population genetics*, concerned with the distribution of genes in different populations; and (6) *genetic mapping*, concerned with the sequences of bases and genes in the chromosomes.

**Genome.** All the genes in a complete set of chromosomes for a species. It is the set of biological instructions for how the individual is formed in the first place and how it later functions. The human genome consists of about forty thousand genes, half of them coming from each parent. Except for identical twins, the gene structure is unique in each individual. The human genome consists of an estimated 3 billion base pairs (6 billion bases in double-stranded DNA). Stretched lengthwise, a single human genome would be about 1.5 meters long. All the genomes in the human body would stretch to the Moon and back about eight hundred times.

**Gravitation.** See **gravity**.

**Gravity.** Also called gravitation, the attraction for one another of two bodies, on account of their mass. The most familiar example is the gravitational force attracting us toward Earth, which we experience as our weight. According to Newton's law of gravity, the force of attraction between two bodies is proportional to the product of their masses and inversely proportional to the square of the distance between them (i.e., the inverse square law applies). Gravity is the weakest of all the known forces. In the hydrogen atom, for example, the gravitational attraction between the proton and the electron is much smaller, by a factor of over $10^{39}$, than the electrical attraction.

**Half-life.** When a nucleus undergoes radioactive disintegration, however much of it we have at a given time, half of that amount will have disappeared after a certain time (refer to fig. 7, p. 44). This time is known as the half-life. This type of decay is known as exponential decay.

**Heavy water.** A substance composed of molecules, $D_2O$, in which an oxygen atom is combined with two deuterium atoms, ${}^{2}_{1}H$ (instead of with two atoms of ordinary hydrogen, as in $H_2O$). It is used in some nuclear reactors.

**Helium.** The element whose atoms have two protons in their nuclei. The two forms of helium, which we write as ${}^{3}_{2}He$ and ${}^{4}_{2}He$, are isotopes of helium. ${}^{4}_{2}He$ is about four times as heavy as an ordinary hydrogen atom, whereas ${}^{3}_{2}He$ is three times as heavy. The proportion of ${}^{4}_{2}He$ found in nature relative to ${}_{2}{}^{3}He$ is 99.99986 percent; the fraction of ${}^{3}_{2}He$ is exceedingly small. Helium is a gas under ordinary conditions.

**Heredity.** The transmission of physical or mental characteristics from one generation to another.

**Hypothesis.** A theory that is not firmly established and is likely to be modified or overthrown. Hypotheses are of great value in science, because testing them often leads to new and significant information. It is better to have a hypothesis that proves to be wrong than no hypothesis at all.

**Indeterminacy principle.** See **uncertainty principle.**

**Information.** That of which we are informed. It comprises perceived facts that are believed to be true, misinformation, and meaningless statements.

**Inheritance.** The transmission of information from parent to offspring by means of genes. The only characteristics that can be transmitted genetically are those encoded in DNA. Traits acquired during the lifetime of the parent, such as muscle development, are not transmitted genetically. Exceptions are certain gross changes that might be brought about by high-energy radiation such as x-rays or radiation from radioactive substances.

**Interference.** Two waves may augment each other, which means that crests coincide with crests, and troughs with troughs, in which case we have what is called constructive interference. Alternatively, troughs may coincide with crests, in which case there is destructive interference.

**Interferometer.** An instrument that exploits interference to measure distances or to make accurate measurements of wavelengths of radiation.

**Inverse square law.** When this law applies to an action involving distance, the effect is inversely proportional to the square of the distance. If the distance is doubled, the effect becomes one-quarter of what it was.

**Isotope.** Atoms whose nuclei have the same number of protons but that differ in the number of neutrons are isotopes of a particular element. Different isotopes of an element are always very similar in chemical and physical properties, which depend much more on the charge on the nucleus than on the mass.

**Judicial method.** The selection of information in an unbiased way in order to give rise to a coherent (self-consistent) conclusion.

**Kinetic energy.** The energy something possesses by virtue of its *motion*. The kinetic energy of a body of mass $m$ moving with a speed of $v$ is $\frac{1}{2}mv^2$.

**Laser.** A device that produces coherent light, which means that the waves are in step with one another, their troughs and crests all coinciding in space and time. Light from a laser is typically monochromatic (i.e., it has a sharply defined wavelength) and may be very intense.

**Law.** In science, this term means different things. It sometimes means an **empirical law**, and sometimes a **universal law**.

**Mutation.** A change in the genetic makeup of a cell or a virus. Mutations may be extensive, such as the loss or modification of a large section of a chromosome. They may be simply the removal or replacement of a single base in a DNA molecule, producing a change in the amino acid sequence in a protein molecule synthesized under the action of the gene. Mutations occur during cell division and may result from a modification of the process of division or as a copying error during the replication of a DNA sequence. If they occur in the cells involved in reproduction they will be inherited and will lead to a modified feature in the offspring. A genetic modification in a single base sometimes results in significant, and usually deleterious, changes in the offspring. Occasionally mutations have a favorable effect on the offspring, and this is an important factor in evolutionary improvement.

**Natural selection.** A crucial factor in biological evolution. Individual members of a species have various chances of survival, determined by their genetic make-up. Those with characteristics that increase their chance of survival are more likely to produce offspring than those with lower chances. As a result, the survival capacity of a species tends to improve over the generations. This is the essential factor in the evolutionary process.

**Neo-Darwinism.** The modern development of the theory of evolution in terms of genes.

**Neutrino.** A class of elementary particle having zero charge and a very small mass, which perhaps is zero. All are very difficult to detect, passing at high speed through matter with virtually no interaction.

**Neutron.** One of the fundamental building blocks of nature. It has approximately the same mass as the proton but carries no electric charge. It is stable when associated with protons in the nucleus of an atom but unstable when isolated, emitting an electron and forming a proton, with a half-life of 15.3 minutes.

**Neutron star.** A highly compressed star consisting of densely packed neutrons held together by gravitational attraction. It typically has a size similar to that of Earth but a mass like that of the Sun. Neutron stars are formed by the gravitational attraction of a massive star that has become a supernova. A pulsar is a form of neutron star.

**Nucleon.** Either a proton or a neutron.

**Nucleotide.** A long-chain molecule consisting of an organic base, a phosphate group, and a sugar.

**Nucleus.** (1) In physics and chemistry, the central part of an atom, made up of protons and neutrons. It is tiny compared with the size of the entire atom. (2) In biology, the part of a living cell that contains the chromosomes.

**Ohm's law.** According to this law, the current that flows through a wire is equal to the voltage drop across the wire divided by the resistance of the wire. Thus if the voltage drop is 10 volts, and the resistance is 5 ohms, the current is $10/5 = 2$ amperes.

**Perpetual motion.** Motion that, if it occurred (it does not), would be in violation of the first or the second law of thermodynamics (or both). Perpetual motion of the first kind is motion that would be possible only if energy could be created from nothing. Perpetual motion of the second kind would violate the second law. One way of looking at this situation is to say that the second law requires that for a process to occur there must be dissipation of a certain amount of heat (this amount can be calculated); a machine that would not dissipate enough heat is impossible according to the second law. An equivalent statement is that such a machine would be impossible if it does not involve an increase in the entropy of the system plus its surroundings.

**Phase space.** An abstract space, points in which relate to the state of a dynamic system.

**Photoelectric effect.** The ejection of an electron from an atom by a photon of sufficient energy.

**Photon.** The tiny packets of which light is composed.

**Plate tectonics.** The branch of geology that deals with movements of Earth's crust on a large scale. The lithosphere, formed by the solid crust of the planet, consists of large, tightly fitting plates that float on the semimolten layer of Earth's mantle. The plates move relative to one another, and over the course of geological time this movement has brought the landmasses into the present arrangement of continents and oceans. The theory of plate tectonics is an expanded and modified version of earlier ideas about continental drift.

**Positron.** The antiparticle of the electron, having the same mass but the opposite charge.

**Protein.** A molecule made up of long chains of amino acids. Proteins are important constituents of all living systems. Meat, fish, and other foods such as beans are important constituents of animal diets in that they provide a source of amino acids for synthesizing proteins.

**Proton.** One of the important building blocks of nature and a constituent of every atomic nucleus. We arbitrarily regard its charge as positive and say that it is a single positive charge; in other words, the charge is +1. This charge is equal to that on the electron but of opposite sign. The number of protons in a nucleus distinguishes one chemical element from another. Protons appear to be infinitely stable. According to modern theory they are composed of quarks.

**Pulsar.** A rotating neutron star that emits pulses of electromagnetic radiation at regular intervals, typically every few seconds.

**Quantum theory.** According to this theory, energy comes in packets. The permitted energy levels are so close together that we usually do not notice the quantization, which only becomes apparent when we deal with systems at the atomic level. When Max Planck suggested quantization in 1900 it was applied to hot radiating bodies and little else. In 1905 Einstein suggested that it also applied to radiation and that it explained other effects. In 1913 Niels Bohr used the idea of quantization to explain the structure and behavior of atoms and molecules, and then it was realized that the theory was fundamental to all of science. In the 1920s the original ("classical") quantum theory was greatly generalized with the development of the new science of quantum mechanics.

**Quarks.** An elementary particle that is a constituent of certain subatomic particles, such as protons and neutrons (refer to chap. 6).

**Quasar.** An astronomical object that when observed optically appears to be a star and that occurs only at enormous distances from Earth. Quasars were first detected by the radio waves that they emit. Since quasars are many billions of light-years away, we see them as they were many billions of years ago. They are believed to be younger galaxies, emitted at the big bang at such enormous speeds (perhaps at 98 percent of the speed of light) that they traveled a great distance in a few billion years.

**Radiation.** Term applied to high-energy particles (such as $\alpha$- and $\beta$-particles) and also to electromagnetic radiation, which can be regarded as composed of a stream of quanta of radiation called photons. The term is also applied to the process of emission of the particles or the waves.

**Radioactivity.** The spontaneous disintegration of an atomic nucleus. High-energy particles, including $\alpha$- and $\beta$-particles as well as electromagnetic radiation ($\gamma$-rays), are emitted in the processes. Radioactivity was discovered in 1896 by French physicist Antoine-Henri Becquerel, whose work was done with uranium, the most abundant radioactive element in nature.

**Radiocarbon dating.** A technique for the dating of archaeological artifacts of organic origin by use of the radioactivity of an isotope of carbon.

**Redshift.** The Doppler shift brought about when an object is moving away from the observer. The name results from the fact that blue light is shifted toward the red, but the term is rather misleading, since it is not simply related to the color of a star.

**Relativity.** The theory that describes how different observers in different frames of reference record the same phenomena. Einstein's special theory of relativity was a dynamic theory that is consistent with the universal constancy of the speed of light and that revealed a relationship between mass and energy. His later, general theory gave a comprehensive treatment of gravitation.

**Replication.** Copying, simply put. A biological replicator has the following characteristics: (1) the form and behavior of the parent are copied (heredity); (2) they are sometimes copied with errors or other variations; and (3) only some features are successfully copied (selectivity). The term *replication* is particularly used in reference to the synthesis of two identical double-stranded DNA molecules from one original double strand.

**RNA (ribonucleic acid).** A molecule that differs from DNA in having a different sugar (ribose instead of deoxyribose) and having the base uracil instead of thymine. RNA occurs in different forms with a variety of functions. For example, messenger RNA (mRNA) is copied from RNA in the nucleus and transmits encoded information in order to provide instructions for the synthesis of proteins.

**Second law of thermodynamics.** This law can be stated in two different but equivalent ways. (1) Kelvin's definition: when any process occurs spontaneously, there must always be some dissipation (wastage) of heat; the minimum amount that must be lost can be calculated for individual systems from the temperatures involved. (2) Clausius's definition: for any process to occur spontaneously, there must be an increase of entropy.

**Selectivity.** In genetics, refers to the fact that sometimes not all features are successfully copied in replication.

**Spectrum.** The range of frequencies (and therefore wavelengths) of radiation. The spectrum extends widely in both directions beyond the narrow range visible to the human eye. In the lower-frequency range, beyond the red, are the infrared, the microwave, and then the radiation used for television and radio transmission. On the other side of the visible, with higher frequencies, are the ultraviolet, x-rays, and gamma rays. Since frequency and energy are closely related we also speak of an energy spectrum.

**Supernova (*pl.*, supernovae).** An enormous explosion marking the final stage of the evolution of a star. A common type of supernova occurs when the core of a massive star has burned all of its nuclear fuel and can no longer resist gravitational forces. It then collapses to a neutron star or a black hole, releasing huge amounts of gravitational energy, comparable to the amount of light given out by an entire galaxy. A recent supernova, visible to the naked eye, was observed in 1987 and is designated Supernova 1987A; it lasted more than a hundred days and was ten thousand times brighter than the brightest star. The importance of supernovae is that without them, life on Earth would not have been possible. Supernovae occurring before our solar system was born threw out the mix of heavier atoms such as carbon and iron that compose our Earth and make life possible on it.

**Theory.** In science, a rational and coherent system of ideas that explains information obtained by observation and experiment. Some theories, such as the theory of relativity and the theory of evolution, are so strongly and universally established that they are called universal laws. Less well-established theories are called hypotheses.

**Thermodynamics.** The branch of physics that deals with the relationship between heat and the other forms of energy. See also the **first law** and **second law**. There is a third law, relating to heat transfer at temperatures close to absolute zero, but it need not concern us here.

**Transform fault.** A phenomenon occurring when plates in Earth's crust slide past one another. The theory of the transform fault was proposed in 1963 by J. Tuzo Wilson as part of the theory of **plate tectonics**.

**Tritium.** The isotope of hydrogen whose nucleus consists of one proton and two neutrons, meaning that it contains three particles and that its mass is three units. Tritium is given the symbol T or ${}^{3}_{1}H$. This isotope of hydrogen is radioactive, with a half-life of 12.3 years.

**Uncertainty principle.** Also called the indeterminacy principle, first stated by Werner Heisenberg, according to which it is *impossible to determine both the position and speed of a particle exactly*. The more accurately one knows the position, the less accurately one knows its speed, and vice versa.

**Universal law.** A scientific principle of great generality that, as far as we know, applies throughout the universe. One of them is the law of gravity, discovered by Newton. Newton's three laws of motion are also universal (or cosmological) laws, as are the first and second law of thermodynamics. Some universal laws are usually referred to as theories: examples are the atomic theory, Einstein's theory of relativity, electromagnetic theory, and the quantum theory. Some important laws and principles, such as Boyle's law and the theory of evolution, are not universal laws but follow logically and inevitably from them.

**Variation.** In a replication process, refers to the fact that characteristics are sometimes copied with errors or changes arising from other causes. The term is also used to mean the range of differences among individuals in a biological population.

**Virus.** An infective agent that is able to multiply only within the living cells of a host. It usually consists of a nucleic acid molecule in a protein coat, sometimes with a surrounding complex of fat and protein. Viruses are generally considered to be nonliving, since they do not survive for long unless attached to a host; they replicate by taking over the host cell's metabolic processes. Many viruses produce pathological conditions, since they can produce serious cellular dysfunction. Examples of diseases caused by viruses are measles, smallpox, poliomyelitis, and various forms of cancer. Viruses are resistant to antibiotics.

**White dwarf.** A collapsed remnant of a star. It is white because it is still glowing and white-hot, and it is considered a dwarf because it is much smaller than the star it was originally, but just as massive.

# SUGGESTED READING

## Scientific Topics

Atkins, Peter. *Creation Revisited*. Oxford: Oxford University Press, 1992.
———. *Galileo's Finger: The Ten Great Ideas of Science*. Oxford: Oxford University Press, 2003.
Davies, Paul. *The Fifth Miracle: The Search for the Origin and Meaning of Life*. New York: Simon and Schuster, 1999.
Dawkins, Richard. *The Selfish Gene*. Oxford: Oxford University Press, 1976; Oxford Paperbacks, 1978, 1989.
———. *The Blind Watchmaker*. London: Longmans, 1986; London: Penguin Books, 1988, reprinted with an appendix, 1991.
———. *River out of Eden*. London: Weidenfeld and Nicholson, 1995.
———. *Climbing Mount Improbable*. New York: W. W. Norton, 1996.
———. *Unweaving the Rainbow: Science, Delusion, and the Appetite for Wonder*. London: Allan Lane, Penguin Press, 1998.
Ekeland, Ivar. *Mathematics and the Unexpected*. Chicago: University of Chicago Press, 1988. On p. 68 of this book is quoted the reference for the effect of gravity on a billiard carom: M. Berry, "Regular and Irregular Motion," in *Topics in Non-Linear Dynamics*, American Institute of Physics Conference Proceedings 46 (N.p.: American Institute of Physics, 1978), pp. 111–12.
Gleick, James. *Chaos: Making a New Science*. New York and London: Penguin, 1987.
Gorst, Martin. *Measuring Eternity: The Search for the Beginning of Time*. New York: Broadway Books, 2001.

Gratzer, Walter. *Eurekas and Euphorias: The Oxford Book of Scientific Anecdotes*. Oxford: Oxford University Press, 2002.

Gribbin, John. *Almost Everyone's Guide to Science: The Universe, Life, and Everything*. New Haven, CT: Yale University Press, 1998.

———. *The Birth of Time: How Astronomers Measured the Age of the Universe*. London: Weidenfeld and Nicholson, 1999; New Haven, CT: Yale University Press, 2000.

———. *Q Is For Quantum: An Encyclopedia of Particle Physics*. New York: Simon and Schuster, Touchstone, 1999.

Hall, Nina, ed. *The New Scientist Guide to Chaos*. New York and London: Penguin, 1992.

Hogan, Craig J. *The Little Book of the Big Bang: A Cosmic Primer*. New York: Springer-Verlag, Copernicus, 1998.

Jones, Steve. *The Language of Genes*. London: HarperCollins, 1993; paperback edition, 1995.

Laidler, Keith J. *To Light Such a Candle*. Oxford: Oxford University Press, 1998.

———. *Energy and the Unexpected*. Oxford: Oxford University Press, 2003.

———. *Science and Sensibility: The Elegant Logic of the Universe*. Amherst, NY: Prometheus Books, 2004.

Leff, Henry S., and Andrew F. Rex. *Maxwell's Demon*. Princeton, NJ: Princeton University Press, 1990.

Lindley, David. *Boltzmann's Atom: The Great Debate That Launched a Revolution in Physics*. New York: Free Press, 2001.

Maddox, John. *What Remains to Be Discovered: Mapping the Secrets of the Universe, the Origins of Life, and the Future of the Human Race*. New York: Simon and Schuster, Free Press, 1998.

Rees, Martin. *Before the Beginning: Our Universe and Others*. London: Touchstone, 1997.

Reeves, Hubert. *Atoms of Silence*. Cambridge, MA: MIT Press paperback, 1985; Toronto: Stoddart, 1993.

Reeves, Hubert, Joel de Rosnay, Yves Coppens, and Dominic Simonnet. *Origins: Cosmos, Earth, and Mankind*. New York: Arcade, 1998.

Schopf, J. William. *Cradle of Life: Discovery of Earth's Earliest Fossils*. Princeton, NJ: Princeton University Press, 1999.

Silk, Joseph. *The Big Bang*. New York: W. H. Freeman, 1989.

Tanford, Charles. *Ben Franklin Stilled the Waves: An Informal History of Pouring Oil on Water with Reflections on the Ups and Downs of Scientific Life in General*. Durham, NC, and London: Duke University Press, 1989.

Thomas, John Meurig. *Michael Faraday and the Royal Institution: The Genius of Man and Place*. Bristol, UK, Philadelphia, and New York: Adam Hilger, 1991.

Thuan, Trinh Xuan. *Chaos and Harmony: Perspectives on Scientific Revolutions of the Twentieth Century*. Oxford and New York: Oxford University Press, 2001.

Von Baeyer, Hans Christian, *Maxwell's Demon: Why Warmth Disperses and Time Passes*. New York: Random House, 1998. Paperback edition, *Warmth Disperses and Time Passes: The History of Heat* (New York: Modern Library, 1999).

Ward, Peter D., and Donald Brownlee. *Rare Earth: Why Complex Life Is Uncommon in the Universe*. New York: Copernicus, 2000.

Weinberg, Steven. *The First Three Minutes: A Modern View of the Origin of the Universe*. New York: Basic Books, 1998. This reprint of a classic first published in 1977 contains an afterword written in 1993, bringing it up to date.

White, Michael, and John Gribbin. *Einstein: A Life in Science*. London: Simon and Schuster, 1993.

Wills, Christopher, and Jeffrey Bada. *The Spark of Life: Darwin and the Primeval Soup*. Cambridge, MA: Perseus, 2000.

Winchester, Simon. *Krakatoa: The Day the Earth Exploded, August 27, 1883*. New York: HarperCollins, 2003. This book gives a particularly readable account of many aspects of geology, particularly plate techtonics.

## BIOGRAPHIES

The reading of biographies of distinguished scientists is an excellent and interesting way of gaining an understanding of science. Biographies are usually written in such a way as to be easily intelligible to readers not trained in science. The following are some biographies that relate to the subject matter of this book.

**Francis Bacon.** The standard biography, and an excellent one, is Catherine Drinker Bowen, *Francis Bacon: The Temper of a Man* (Boston and Toronto: Little, Brown, 1963).

**Niels Bohr.** Abraham Pais, *Niels Bohr's Times* (Oxford: Oxford University Press, 1991).

**William Buckland.** W. F. Cannon, *Dictionary of Scientific Biography* (New York: Scribner, 1970), 2:566; N. A. Rupke, *The Great Chain of History: William Buckland and the English School of Geology* (Oxford: Clarendon Press, 1983).

**James Chadwick.** Andrew Brown, *The Neutron and the Bomb: A Biography of Sir James Chadwick* (Oxford: Oxford University Press, 1997).

**Charles Darwin.** Adrian Desmond and James Moore, *Darwin: The Life of a Tormented Evolutionist* (New York and London: Norton, 1991). A shorter biography, also excellent, is Michael White and John Gribbin, *Darwin: A Life in Science* (New York: Simon and Schuster, 1995).

**Albert Einstein.** Abraham Pais, *Subtle Is the Lord: The Science and the Life of Albert Einstein* (Oxford: Clarendon Press, 1982). A biography that deals more with personal aspects is Roger Highfield and Paul Carter, *The Private Lives of Albert Einstein* (New York: St. Martin's Press, 1993). Other biographies that are interesting and explain the physics well are Michael White and John Gribbin, *Einstein: A Life in Science* (London: Simon and Schuster, 1993, paperback edition, 1994); and Denis Brian, *Einstein: A Life* (New York: Wiley, 1996).

**Benjamin Franklin.** The standard biography is Carl Van Doren, *Benjamin Franklin* (New York: Viking, 1938). A more recent one is Walter Isaacson, *Benjamin Franklin* (New York: Simon and Schuster, 2003). An account particularly of Franklin's scientific work on electricity is Tom Tucker, *Bolt of Fate: Benjamin Franklin and His Electronic Kite Hoax* (New York: Public Affairs, 2003). The account is an interesting one, but many of the conclusions are rather speculative.

**Rosalind Franklin.** Brenda Maddox, *Rosalind Franklin: The Dark Lady of DNA* (London: HarperCollins, 2002), is a particularly admirable biography that explains in an easily understood way the work on the structure of DNA and the resulting controversy about priority.

**Galileo Galilei.** There are many biographies of Galileo. A good fairly recent one is Pietro Redondi, *Galileo Heretic* (Princeton, NJ: Princeton University Press, 1983). This book presents the point of view that Galileo was disciplined by his church not so much for his advocacy of the Copernican system as for his theological opinions.

**Werner Heisenberg.** David C. Cassidy, *Uncertainty: the Life and Science of Werner Heisenberg* (New York: W. H. Freeman, 1991). Besides giving a good account of the science this book analyzes in some detail Heisenberg's ambiguous and controversial relationship with the Nazi regime.

**Lord Kelvin.** J. G. Crowther, *British Scientists of the Nineteenth Century*

(London: Kegan Paul, Trench & Trubner, 1935), includes an excellent article on Kelvin. Another valuable account is in D. K. C. Macdonald, *Faraday, Maxwell, and Kelvin* (New York: Doubleday, 1954). There is also an excellent, detailed biography by C. W. Smith and M. N. Wise, *Energy and Empire: A Biographical Study of Lord Kelvin* (Cambridge: Cambridge University Press, 1989).

**G. N. Lewis.** Lewis's son Edward S. Lewis has published *A Biography of Distinguished Scientist Gilbert Newton Lewis* (Lewiston, NY: Edwin Mellen Press, 1998), which gives a clear account of Lewis's main contributions.

**James Clerk Maxwell.** The first biography of Maxwell was by Lewis Campbell and William Garnett, *The Life of James Clerk Maxwell* (London, 1882; reprint, New York: Johnson Reprint, 1969). This book is a gold mine of valuable and interesting information. More recent biographies of Maxwell are C. W. F. Everitt, *James Clerk Maxwell: Physicist and Natural Philosopher* (New York, Scribner, 1975); Ivan Tolstoy, *James Clerk Maxwell: A Biography* (Chicago: University of Chicago Press, 1981); M. Goldman, *The Demon in the Aether: The Story of James Clerk Maxwell* (Edinburgh: Paul Harris, 1982); D. K. C. Macdonald, *Faraday, Maxwell, and Kelvin* (New York: Doubleday, 1954).

**Gregor Mendel.** Robin Marantz Henig, *The Monk in the Garden: The Lost and Found Genius of Gregor Mendel, the Father of Genetics* (Boston: Houghton Mifflin, 2000).

**Isaac Newton.** There are very many biographies of Newton; I have found the following particularly helpful: R. S. Westfall, *Never at Rest: A Biography of Isaac Newton* (Cambridge: Cambridge University Press, 1980); and A. Rupert Hall, *Isaac Newton, Adventurer in Thought* (Cambridge: Cambridge University Press, 1992).

**Count Rumford.** For a short biography see S. C. Brown, *Dictionary of Scientific Biography* (New York: Scribner, 1976), 13:350–52. An excellent full biography is G. I. Brown, *Count Rumford: The Extraordinary Life of a Scientific Genius* (Stroud, Gloucestershire: Sutton, 1999). Some amusing details about Rumford, particularly with reference to his part in the founding of the Royal Institution, are to be found in John Meurig Thomas, *Michael Faraday and the Royal Institution: The Genius of Man and Place* (Bristol, UK, Philadelphia, and New York: Adam Hilger, 1991); and in two of Thomas's recent articles: "Sir William Thompson, Count Rumford and the Royal Institution," *Notes and Records of the Royal Society* 53 (1999): 11–25; and "Rumford's Remarkable Creation," *Proceedings of the American Philosophical Society* 142 (1998): 597–613.

**Ernest Rutherford.** An excellent recent biography is John Campbell, *Rutherford: Scientist Supreme* (Christchurch, New Zealand: AAS Publications, 1999). Several previous biographies are referred to in this book.

**Erwin Schrödinger.** Walter Moore, *Schrödinger: Life and Thought* (Cambridge: Cambridge University Press, 1989), is a particularly excellent biography, covering both the life and the science.

**J. J. Thomson.** An interesting short biography, with helpful illustrations, that presents the science clearly is E. A. Davis and I. J. Falconer, *J. J. Thomson and the Discovery of the Electron* (London: Taylor and Francis, 1997). There is also a good biography by G. P. Thomson, J. J.'s son, and himself a Nobel Prize winner in physics: *J. J. Thomson and the Cavendish Laboratory in His Day* (Cambridge: Cambridge University Press, 1965).

# INDEX

Page numbers appearing in *italics* refer to illustrations or tables.